The Art of the Natural Home

Rebecca Sullivan is an ethical food and agricultural academic, activist, writer, curator, teacher at Le Cordon Bleu Australia and a home cook. A social entrepeneur, as co-founder of well-being brand Warndu, she is passionate about growing food, homesteading, heritage, preserving traditions and passing on a wealth of 'granny skills' to future generations. Her first book, *Like Grandma Used to Make*, was published in Australia.
@grannyskills (instagram)

The Art of the Natural Home

A ROOM BY ROOM GUIDE

Rebecca Sullivan

photography by Nassima Rothacker

KYLE BOOKS

For Mum

First published in Great Britain in 2017 by
Kyle Books, an imprint of Kyle Cathie Ltd
192–198 Vauxhall Bridge Road
London SW1V 1DX
general.enquiries@kylebooks.com
www.kylebooks.co.uk

10 9 8 7 6 5 4 3 2

ISBN 978 0 85783 406 5

Project Editor: Tara O'Sullivan
Editorial Assistant: Isabel Gonzalez-Prendergast
Copy Editor: Stephanie Evans
Designer: Laura Woussen
Photographer: Nassima Rothacker
Illustrator: Juliet Sulejmani
Stylists: Rachel de Thample and Rebecca Sullivan
Prop Stylist: Jennifer Haslam
Production: Lisa Pinnell

A Cataloguing in Publication record for this title is available from the British Library.

Colour reproduction by ALTA London
Printed and bound in China by C&C Offset Printing Co., Ltd.

Important note:
The information and advice contained in this book are intended as a general guide to using plants and are not specific to individuals or their particular circumstances. Many plant substances, whether sold as foods or as medicines and used externally or internally, can cause an allergic reaction in some people. Neither the author nor the publishers can be held responsible for claims arising from the inappropriate use of any remedy or healing regime. Do not attempt self-diagnosis or self-treatment for serious or long-term conditions before consulting a medical professional or qualified practitioner. Do not undertake any self-treatment while taking other prescribed drugs or receiving therapy without first seeking professional guidance. Always seek medical advice if any symptoms persist.

Contents

Introduction 6

Home

Home & Cleaning 12

The Kitchen 32

The Drinks Trolley 94

The Garden 106

Health & Beauty

Face & Body 124

Hair & Make-up 146

For Men 160

Remedies 168

Resources 187

Index 188

Introduction

These days, more than ever, we appreciate the importance of health and well-being, and place great emphasis on the word 'natural'. For many of us, though, this is limited to focusing on what we eat and drink. If we really want to achieve a natural lifestyle and optimum health and well-being, we need to take a more holistic approach. And that is what this book is about – natural recipes and products for the entire home, from the kitchen pantry to the cleaning cupboard, from your make-up bag to the medicine cabinet (and, happily, a few treats for the drinks trolley too).

Of course, food is a key part of this philosophy. For the best part of a decade, my work has involved campaigning and teaching people about what foods we put into our body. From a sustainability perspective I am concerned with food ethics, not just environmentally but also socially and economically. Trying to be a conscious consumer can feel like an uphill battle at times. Reading nutrition labels is like deciphering the Da Vinci Code and they can be confusing and often misleading. Personally, I prefer to avoid fads or trends. Instead, I believe whole-heartedly in simply eating local, seasonal food and knowing where it comes from. It is more important for me to know the farmers and how they raise their animals or tend to their land than to subscribe to a particular label. For me it is quite simple: it is about eating like our grandparents would.

Back in the day (you know, the good old days, before processed foods) they ate local and seasonal food and never wasted a thing. Not because they were hipsters or because it was trendy – no, it was because, back then, local and seasonal foods were cheaper and more available. Waste was just stupid and, unlike today, nobody did it. It was

uneconomical and made no sense. People back then had to cook from scratch (even if they weren't naturally gifted cooks) and make use of everything they could. Sadly, as our lifestyles have changed, this hands-on approach to life is becoming less common, and something important is being lost.

For me, a vital part of my journey towards this lifestyle was my great grandma Lil. In her day, she was an award-winning baker and yet I never saw her cook. When she passed at the age of 100, my mum gave me some of her things and among them were medals for her Victoria sponges. I was heartbroken that I hadn't learned to cook from her, and so it began – a mission to save these life skills and recipes from being forgotten, and to make sure that the wisdom of our grandparents isn't lost. It might have started with that Victoria sponge recipe, but soon it stretched beyond the kitchen and into the entire home.

I realised that, if I was going to bang on about what I put into my body from a food perspective, I should also be paying attention to what I was putting *on* my body. It's the same body, after all. Lil only ever used natural soap – that was it. My inner granny raged when I discovered the true nature of the skincare products I had been using for so many years – products that I thought were natural but in truth were far from it. Once again, the wool had been pulled over my eyes thanks to the language of labels. I was determined to follow Lil's example and make sure that I was applying a natural approach to my skincare and beauty routine too.

This quickly led to me learning how to make my own home-cleaning products, as well. What is the point in religiously eating natural ingredients, only to wash our dishes and clothes and douse our homes in chemical-laden liquids and sprays?

'Natural' can mean a great many things, and as a word it is often overused. To me, calling a product natural should, more often than not, mean that you can eat it with no repercussions. Ninety per cent of the ingredients in this book are edible.

The recipes and methods I have presented here have not just been inspired by generations past. Some have stemmed from my career in developing recipes and my masters in sustainable agriculture, as well as my recent study of herbal medicine basics. This gave me a better understanding of what herbs and flowers, as well as other ingredients, can do for our health and well-being. If you turn to pages 168–186, you'll find my collection of Remedies – natural tinctures and tonics, just like the ones that have been used as commonplace medicines for centuries.

Throughout the book, I have intentionally created recipes from things you can easily grow at home (either in a garden or even on a balcony in a herb box) or indeed forage for. Foraging is all the rage, it's true, but picking things you are not familiar with can be dangerous and also selfish if you get greedy. Remember to be polite, take a little, not a lot and leave some for the birds, bees and other biodiversity that considers it their home. In addition, all of the ingredients I've used are readily available in stores and online, so you should have no problem sourcing them and recreating these recipes and remedies in your own home.

I hope you will get as much joy as I do from exploring the multiple uses that can be found for the most humble of ingredients. Lemons can be used for everything from air-freshener sprays to natural hair-lightening treatments, as well as a truly delicious Citroncello for your cocktail counter. Salt becomes not only a seasoning but also a fabric softener, a beautiful body scrub and a rehydrating face spray. Vinegar works double time as a carpet cleaner, a medicine and a wonderful flavouring when infused with garlic, chillies, herbs or fruit. Honey cleanses skin as a facial toner, sweetens homemade ice creams and sorbets, and soothes sore throats as a cough syrup.

So take a leaf out of our grandparents' books, and create your own beautiful, natural home.

Storecupboard Essentials

- Vinegar (white and apple cider)
- Salt (coarse and non-iodised, as well as some flaked, Epsom and magnesium flakes)
- Bicarbonate of soda
- Castile soap (natural and available online or in most good stores)
- Citrus fruit, peels and juice
- Jars and spray bottles
- A couple of different essential oils, such as lavender, peppermint and eucalyptus
- Microfibre or good-quality dishcloths
- A good scrubbing brush
- Cheesecloth and muslin cloth
- Coconut sugar or raw sugar
- Nut oils (almond, jojoba, coconut)
- Olive oil
- Garlic (a LOT of garlic!)
- Dried herbs and flowers, such as sage, holy basil, rose petals, chickweed, hibiscus, rose, thyme, calendula, lavender, rosemary
- Honey (raw and local)
- Twine
- Scissors
- Pestle and mortar
- Natural colouring (for soaps and candles)
- Beeswax, soy or paraffin wax
- Candle wicks
- Mica powders

Home

WHEN I WAS YOUNGER, I never understood why people loved spending so much time at home. I was a woman (OK, a girl) filled with wanderlust – a desire to be anywhere but home. Travel made me feel alive and inspired. These days I travel a lot for work and, while it still makes me feel alive and inspired, the difference now is that I cannot wait to be home. I love to be in my own bed, with my own kitchen to cook in and my own bathroom, filled with soaps and scrubs that have not been lugged halfway across the world in a suitcase, or accidentally left behind in hotel rooms and showers. I love to be in my warm, happy and clean space. Not sterilised, mind, just clean – and there is a big difference.

My mum is a cleaner. Not by way of profession – she just loves to clean. Ever since I can remember, my siblings and I have teased her about her obsession with cleaning. Sadly, back when we were kids, cleaning meant bleach, antibacterial bottles and aerosol sprays. That is no longer the case; my mum is massively into her eco-cleaning these days. In hindsight I am grateful to have grown up in a clean home and, although when I was a teenager I was certainly NOT a cleaner, I have now turned into my mother. I love to clean (OK, not my bedroom – that remains a disaster) but the communal areas are pretty spotless for the most part. What I really don't like, however, is the array of chemical-laden sprays and scrubs in the supermarkets. Cleaning is big business and, while it's actually wonderful to see the small section of eco- and earth-safe cleaners on the shelves – those are my choice should I have to buy something – my first preference is to go back a couple of generations, before all the chemical products became the norm, to my grandmother's way of cleaning. To her mind, there is nothing you cannot fix with lemon, salt, bicarbonate of soda and vinegar, and now I think the same too. Just as we all want to be sure that the food we eat is keeping us healthy, I want the same natural approach to cleaning my home without harming the environment.

In this chapter, I have collected together my favourite recipes for household products that will give you clean, freshly scented results – without the chemicals.

Home &
Cleaning

Room Sprays

When you don't have fresh flowers but you wish your house smelled like them, try making your own scented spray. Play around with the scents to your liking. Both of these air fresheners will store for up to a year and you may use them as often as you like. For a more intense smell, just add essential oils.

ROSEMARY, SAGE AND LAVENDER AIR FRESHENER

This is an invigorating and floral air freshener – not sickly or fake smelling, more like your nana's garden in spring. Rosemary and lavender go together perfectly.

MAKES 500ML
4 sprigs rosemary
2 sprigs of sage
1 teaspoon lavender
2 lemons, sliced
500ml water

a 500ml spray bottle

Put all the ingredients into a small saucepan. Bring to a simmer over a medium heat, cover, and let it simmer for a further 5 minutes. Let it cool, then strain the liquid and pour it into a spray bottle. You can add a fresh sprig of rosemary to infuse in the bottle, if you wish. If the sprig is too tall for the bottle, just break it in half.

If you like, replace the herbs every couple of weeks as they lose their colour. This will store for up to a year and you can use it as often as needed.

LEMONGRASS, LIME AND GINGER AIR FRESHENER

This one is a great pick-me-up air freshener that will give your home a clean, energising smell.

MAKES 500ML
2 limes, sliced
2 fresh lemongrass stalks
10cm piece of fresh ginger, crushed under a knife
500ml water

a 500ml spray bottle

Put all the ingredients into a small saucepan. Bring to a simmer over a medium heat, cover, and let it simmer for a further 5 minutes. Let it cool, then strain the liquid and pour it into a spray bottle. You can add a fresh lemongrass stalk to infuse in the bottle, if you wish.

Candles

Smell is one of our most powerful senses, with the ability to evoke nostalgia, whet your appetite or instil calm. Everything you need to make these candles is available online, and affordable too, but a lot of this stuff you will have at home. Once you nail this recipe, they will be your go-to gift, as you can personalise the scents.

MAKES ABOUT
8 SMALL CANDLES

2kg soy wax or paraffin
scented or essential oils
mica or candle powders in various
   colours (optional – the amount
   you'll need depends on the
   desired colour, so start with
   ½ teaspoon and go from there)

double boiler (or use a heatproof
   bowl and a saucepan)
thermometer
candle moulds (for example small
   glass jars, tins and pots)
spray oil
wicks
hairdryer

Begin by grating or chopping your wax or paraffin. The smaller you chop it, the quicker it will melt. If you are using a double boiler, get it set up. If you don't have one, boil some water in a large saucepan, then set a heatproof bowl on top of the saucepan. Add the wax to the top of the boiler or the bowl and leave to melt, stirring every so often. Use a thermometer to ensure the temperature of the wax does not exceed 90°C.

Once the wax has melted, add your desired oil or combination of oils (for a 250ml-sized candle, about 30ml of oil is ample) and powder for colour, if you wish. I would do this over the heat, very quickly, so that the colouring agent combines easily. Remove from the heat once everything has been added and mixed so that the colour and oil are distributed evenly. Set aside and quickly prep your mould.

If you're using a temporary mould, spray it with a little oil for easy removal. Next, insert the wick by tying the wick to a pencil and sitting it horizontally across the top of the mould so that the wick hangs vertically. Pour in the wax to about 2cm from the top. The candle sometimes shrinks in the centre, so you can add a little more wax if needed. Use a hair dryer to dispel any air bubbles or divots, and smooth the top.

Cut off the wick and leave the candle for at least 24 hours before removing from the mould or lighting it if using a permanent mould.

Candle colouring is available in every colour
of the rainbow and can be purchased online.

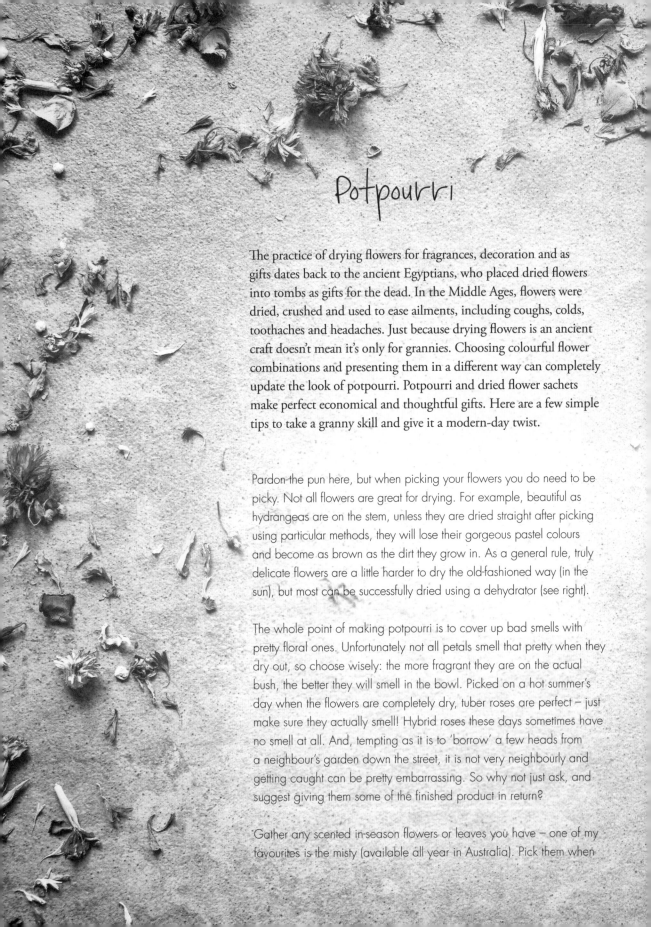

Potpourri

The practice of drying flowers for fragrances, decoration and as gifts dates back to the ancient Egyptians, who placed dried flowers into tombs as gifts for the dead. In the Middle Ages, flowers were dried, crushed and used to ease ailments, including coughs, colds, toothaches and headaches. Just because drying flowers is an ancient craft doesn't mean it's only for grannies. Choosing colourful flower combinations and presenting them in a different way can completely update the look of potpourri. Potpourri and dried flower sachets make perfect economical and thoughtful gifts. Here are a few simple tips to take a granny skill and give it a modern-day twist.

Pardon the pun here, but when picking your flowers you do need to be picky. Not all flowers are great for drying. For example, beautiful as hydrangeas are on the stem, unless they are dried straight after picking using particular methods, they will lose their gorgeous pastel colours and become as brown as the dirt they grow in. As a general rule, truly delicate flowers are a little harder to dry the old-fashioned way (in the sun), but most can be successfully dried using a dehydrator (see right).

The whole point of making potpourri is to cover up bad smells with pretty floral ones. Unfortunately not all petals smell that pretty when they dry out, so choose wisely: the more fragrant they are on the actual bush, the better they will smell in the bowl. Picked on a hot summer's day when the flowers are completely dry, tuber roses are perfect – just make sure they actually smell! Hybrid roses these days sometimes have no smell at all. And, tempting as it is to 'borrow' a few heads from a neighbour's garden down the street, it is not very neighbourly and getting caught can be pretty embarrassing. So why not just ask, and suggest giving them some of the finished product in return?

Gather any scented in-season flowers or leaves you have – one of my favourites is the misty (available all year in Australia). Pick them when

the petals are almost ready to fall off the receptacle (the part the petals are attached to). Before you pick the flower, give it a little shake so that any bugs hiding inside can go and find a new home.

My favourites are rice flowers (they look like little rice puffs), roses, dandelions, violets, lavender, cornflowers, nasturtiums, gerberas, snapdragons, poppies and sage leaves. I always include some flowers more for their looks than their scent, so that the finished mixture looks bright and textured.

Once picked you can either use a dehydrator (see right) or dry the petals out the old-fashioned way. Put some sheets of kitchen paper on trays and lay out petals separately, spreading them apart so the air can circulate. The paper helps to draw any moisture away from the petals. Find a place out of draughts to dry the petals, preferably inside. Near a window is good to give them sufficient sunlight for drying – direct sunlight dries them quicker but they will fade a little.

Leave them for 1–2 days. Check on them as they may dry quicker depending on your home: they like dry heat, no humidity. When they're ready, they feel like tissue paper and wrinkle up a little. Once they are done, I suggest putting them straight into an airtight container lined with kitchen paper towel until you are ready to use them.

If you choose a strong-smelling flower, your potpourri should retain its scent for about six months, and you can add essential oils as the smell dissipates. Choose your mix of oils and add a few to your dried flowers. Try to be sparing to start with, depending on how much mix you have, and add more drops if necessary.

Then place the mix into a pretty jar or tip into a linen bag and voilà, not so granny-like potpourri. I also love to fill a pretty little ceramic or metal bowl with potpourri and place in guest rooms and the bathroom.

USING A DEHYDRATOR

Preheat the dehydrator to a low heat (about 35–45°C, although this will vary.) The thicker the petal, the higher the heat needed, so for roses as an example you may even need to reduce the heat to about 30°C. Before you add them to the dehydrator, remove any stems from the flowers and shake off any critters from the flower back into the garden – I'm not sure they would appreciate the sunburn! Check on your flowers every hour until they are fully dehydrated to your liking. As long as there is no moisture in the petal, your flower will last.

Wood Polish

Here's a great polish to stop your wood from ageing and also to help spruce it up when it's looking tired. This works on dark wood best. I would suggest you test this on a small, inconspicuous area first to see how you like the post-polish look.

Put everything into a spray bottle and shake to mix. This will keep in the cupboard under the sink or in the laundry room for up to a year. When using, spray lightly on the surface and rub with a soft cloth to polish.

MAKES 225ML
3 tablespoons olive oil
180ml vinegar (use white vinegar for
    light wood and apple cider vinegar
    for dark wood)
30 drops orange oil or pine oil, or
    combination of both

1 spray bottle

The Magic of Bicarb

Bicarbonate of soda, or sodium bicarbonate, has been used for centuries for cleaning. It was even used by the ancient Egyptians.

It can be used all over your house to clean almost everything, whether you sprinkle it over surfaces before wiping it away, or turn it into a paste mixed with water to scrub at more stubborn marks. You'll find it used in many of my cleaning recipes.

Soap

Making your own soap is incredibly gratifying. It is lovely knowing that each time you have a shower or bath the lathery bar is filled with herbs and flowers of your choosing to make all your favourite smells. It can be a little daunting making soap from scratch as it usually requires a chemical called lye, but this recipe does away with the need for lye, so it's a great one to start with. This recipe uses instead a natural product known as 'melt and pour soap' as your base. This, along with all of the equipment, is easily available online.

MAKES 2–4 BARS
(depending on size)
500g melt and pour soap base
natural soap colouring
your favourite herbs
   (dried or fresh)
your favourite flowers
   (dried or fresh)
a few drops of your favourite
   essential oils

moulds (flat tins, small jars or
   silicone chocolate moulds are
   all suitable)
double boiler (or use a heatproof
   bowl and a saucepan)
thermometer

Start by chopping up your soap base into small chunks, or grate it, to make it melt faster. Prepare your moulds by spritzing them with a little oil for easy removal afterwards. Organise the colours you want to use and decide on the combinations of herbs, flowers and aromas prior to melting.

Melt the soap base in a double boiler to the temperature recommended by the manufacturer. Once melted, stir in your colour (start with a few drops or pinches until you get the colour you desire). Remove from the heat. Once the mixture has cooled slightly, add 6–10 drops of your favourite essential oil. If you are using more than one oil, blend them together first to ensure you get the scent you desire. Mix through.

Pour your soap into moulds. You can either add flowers/herbs to the bottom and top of the moulds, just sprinkle them over the top, or stir them all the way through. Allow to set for 24 hours in a cool place, but not in the fridge or freezer.

Remove from the moulds when set. If you have trouble removing the soap, put the mould in a bowl of boiling water momentarily until the soap loosens. You can also use a sharp knife to help ease it out.

These make a perfect gift when presented in a jar with a ribbon.

Laundry

WASHING POWDER

My partner, Damien, loves to wash his clothes.
I am more of a 'if it doesn't have a stain or smell
like a bonfire, it ain't getting washed' kind of girl.
Damien really loves clean clothes, so this recipe is
for him. And for anyone who suffers from eczema,
this will ensure you don't get irritated by any
washing powder residue left on your clothes.

MAKES 650G

100–150g bar of soap (real soap like Dr. Bronner's
    or castile, not glycerine filled)
250g bicarbonate of soda
250g washing soda (soda crystals)
5 drops lemon or orange essential oil

Chop the soap into small pieces. Place all of the
ingredients into the food processor and blitz to a fine
powder. Let it settle before opening the lid as the
particles are fine. Use between 1–3 tablespoons,
depending on how dirty your clothes are.

FABRIC SOFTENER

Perfect for your towels and sheets to give them
that just-washed-by-your-granny feel. Salt actually
softens fabric and the lavender smells amazing
on your sheets. You can change the essential oil
if you wish.

MAKES 750G

500g Epsom salts
250g baking soda
30 drops lavender essential oil

Mix all of the ingredients together and store
in an airtight container until ready to use. Put
2 tablespoons in your washing load.

*The sun is a magical thing.
Hanging your sheets to dry in
the sun doesn't just leave them
smelling clean and fresh. The
sun is the perfect safe bleach
– it will lighten anything you
leave in it over time.*

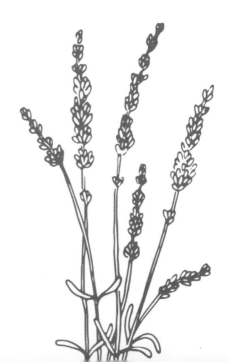

Carpet Cleaner

Vinegar – like lemon juice – is a natural acid and is great at extracting stains from carpets. In this easy remedy I combine it with eucalyptus oil, which is great at removing anything sticky or waxy and also smells wonderful. I recommend you test this cleaner on an inconspicuous patch of carpet, to make sure it works with your carpet fibres, before you use it on the entire floor.

MAKES 750ML
250ml white vinegar
500ml water
3 tablespoons sea salt
20 drops eucalyptus oil

a 750ml spray bottle

Combine the vinegar and water in a spray bottle. Add the salt and oil. Shake well to combine. Store in a cool, dark place for up to three months. Shake before use.

Spray on the carpet and leave overnight. Vacuum the next day. Use when you have a stain by directly spraying the mix onto the stain or spray all over once a month to keep your carpet clean and fresh-smelling.

Moths Be gone

There is nothing more irritating than pulling out your favourite sweater at the beginning of winter to find it has holes in it from being eaten by the moths. Store-bought moth repellents can be full of nasties, such as camphor and dichlorobenzene, and often smell awful. These handmade ones are not nasty and will also make your clothes smell delightful. In short, the moths hate these smells and should stay away from your cashmere.

MAKES 2 SACHETS
a mixture of any of the following herbs, dried (see how to dry herbs on page 34):
½ handful of rosemary and ½ handful of peppermint or garden mint or
½ handful of lavender and ½ handful of rosemary or
$1/3$ handful of dried citrus peel, broken-up cinnamon stick and lavender

an old pair of tights or stockings

Cut off the legs of your stockings or tights – you only need the foot part for this. Mix your herbs together, then pour them into the foot. Tie a knot in the top to close, then finish with a ribbon or twine. Place in your drawer, give it a scrunch and hey presto! Scrunch to release oils every so often (about once a month) to keep the moths at bay. Replace every six months with a fresh batch. If you want to use the leg of your stockings or tights to avoid waste, just tie knots in both ends.

Oven and Hob Cleaning

OVEN CLEANER

There is nothing worse than the smell of chemical oven spray. It literally makes you cough and the labels always warn you to avoid any contact with your skin, so you can only imagine what is in it. Sure, it's quick and easy, but a tiny bit of muscle power and patience and you can have a chemical-free, clean oven.

MAKES ENOUGH FOR A SINGLE USE
60g bicarbonate of soda
a little water
120ml white or apple cider vinegar

Place the bicarbonate of soda in a small bowl and add a little water at a time until it forms a paste. Wearing rubber gloves, take a cloth and rub the paste over the entire surface of your oven. Depending on the size of your oven, you may need to make a little more. Leave for 12 hours to work its magic.

After that time, wipe all of the surfaces with kitchen paper and discard. Put the vinegar in a spray bottle and spritz all surfaces. Use a cloth and some warm water to continue to remove the residue. Give all surfaces a final wipe down with clean warm water and leave to dry.

STOVETOP SPRAY

I would suggest giving your stove a wipe down with this spray every time you use it to avoid build up of grease and grime. I have given ratios rather than quantities so you can make as much as you need.

1 part salt
1 part bicarbonate of soda
1 part water

a spray bottle

Put everything into a spray bottle. Shake well before use. Spray the surface and wipe with a clean, damp cloth. This works best made fresh.

Washing-up Liquid

There are actually lots of fantastic eco washing-up liquids on the market, but I'm always super concerned with what's going down my sink into our waterways, so I gave this recipe a go and it works just as well. I would suggest giving your dishes a little rinse and wearing gloves – as you probably do anyway.

MAKES 450ML

1 tablespoon borax susbtitute or bicarbonate of soda
1 tablespoon castile liquid soap
450ml water, boiled from the kettle
10 drops lemon essential oil

Put the borax susbtitute or bicarbonate of soda and soap in a medium-sized bowl and pour over the boiling water. Mix until it is all combined. Allow the mixture to cool completely; it will form a gel-like consistency. Pour into a storage bottle (something easy to use like a squirt bottle). Add the essential oils, shake and store under the sink. When using, add a little at a time to hot water depending on how soapy you like the water to wash your dishes. This will keep for about one year.

Citrus All-round Cleaning Spray

This will clean and sterilise all surfaces – kitchen and bathroom – and leave everything smelling delicious. This makes a fairly small amount, but you can easily double or triple the ingredients.

MAKES 100ML

peel from 1 citrus fruit (although if you choose limes, use two as they are smaller)
50ml white vinegar
50ml water

a small glass jar
a spray bottle

Tightly pack the citrus peel into a glass jar and cover with white vinegar. Put the lid on, and let it sit for 2–4 weeks. Strain the vinegar into a spray bottle and top up with an equal amount of water. Spray directly onto surfaces and wipe with a damp cloth. This will keep indefinitely.

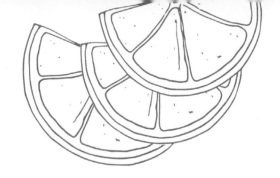

The Bathroom

SPRAY AND WIPE

This is perfect for cleaning every single surface in
your bathroom and hard surfaces in the rest of the
home too. This is the best way to avoid cleaners full
of chemicals and it's easy to make at a fraction of
the cost of bought products, which often contain
endless lists of ingredients on back labels – and even
I don't know what some of them are. Far better to
make your own from scratch – that way you have
full control of what you use in your home. Do make
sure you use a glass bottle for this cleaner.

MAKES 500ML
1 teaspoon borax substitute or bicarbonate of soda
½ teaspoon natural washing soda (soda crystals)
1 teaspoon liquid castile soap
20 drops essential oil
500ml warm filtered water

a 500ml glass spray bottle

Put all ingredients into a sterilised spray bottle. Shake
well and use as needed. Store in a cool, dark place
for up to three months.

*Shop-bought bathroom cleaning products can be among the worst
culprits for chemical ingredients. Make your own at home, and you'll
know exactly what they contain.*

SCUM SCRUB

That stubborn scum that ends up in the cracks of the bathroom tiles or on the bottom of the basin is not the easiest of things to clean because it's made up of various elements – mildew, mineral deposits and the limescale from hard water, to name but a few. Try to prevent the build-up by wiping down the shower with a cloth after your wash. Prevention will avoid too much scrubbing. This product is best used weekly in your bathroom clean. Simply sprinkle onto a damp surface, scrub and rinse. Test on a small, inconspicuous area first to make sure this is suitable for use in your bathroom.

MAKES 500G
250g bicarbonate of soda
125g non-iodised salt
125g natural washing soda (soda crystals)
4–6 drops orange oil

lidded glass jar for storage (if you wish, you can use two lids and poke holes in one for sprinkling, then replace it with the full lid for storage)

Mix all the ingredients together into a bowl. Place into a jar and shake. Use a spoon to sprinkle. Store in a dry, dark and cool place. Use within three months.

Once a week, ever so slightly wet the scummy surface. Sprinkle your scum scrub generously and let it sit for 10 minutes. Scrub with a brush and use an old toothbrush for the tile grout. Rinse.

TOILET BLISS BOMBS

I don't know about you, but I don't like those fluorescent sticky toilet bowl tabs sold in the supermarket. After all, with every flush, those chemicals will end up in your nearest waterway then eventually back through the water from your tap. Ew. If you do want to make sure that your toilet is clean and fresh without using the chemical bowl tabs, here's how. These bliss bombs make the loo both clean and fresh smelling.

MAKES 12–14
250g bicarbonate of soda
80g citric acid
10 drops rose oil
10 drops lavender oil
5 drops lemon oil

spray bottle
rubber gloves
cotton face mask
baking tray lined with baking paper or ice-cube tray
glass jar for storage

Wear gloves and a face mask – citric acid is strong and could make you cough. Mix the bicarbonate of soda and citric acid in a glass bowl. Fill a spray bottle with some water and very gradually add water to the bowl by spraying and mixing, just enough to make the ingredients stick together. Add the essential oils and mix. Press small amounts of the mixture together into small balls, then place on the tray. Let them dry overnight, then remove and store in a glass or other airtight jar container. To use, drop one into the toilet, leave it to dissolve and flush on the next use. Use as necessary but no more than once a day.

The Kitchen

ANYONE WHO KNOWS ME knows that my kitchen is
my happy place. It always has been. Not just my own – anyone's
kitchen, really.

As a child, the kitchen was the centre of our home. Everyone in my
family hung out in or near the kitchen, watching my grandmother
prepare a feast. I didn't know it back then, but she would become
my inspiration and the reason for where I am today. Pauline, my
grandmother, is an incredible cook. She never wastes a thing - the
poster child for sustainable living. Frugal, smart, resourceful. She is
thoughtful and generous and can make a beautiful gift from things
most would throw away.

As you will see throughout this book, this attitude is one that I try
to apply to every aspect of my life – but it is a life that is still centred
around cooking and food. I could fill whole books with recipes
and projects for the kitchen alone, but for now this chapter gives a
selection of essentials, staples and treats for a zero-waste, chemical-
free kitchen and larder.

Saving Herbs

There are three really easy ways to use any leftover herbs so you don't waste them. Herbs like mint, parsley, basil and chives are best used fresh, but you can also freeze them in ice-cube trays. Harder herbs like sage, rosemary and oregano are great dried.

DRYING INDIVIDUAL LEAVES

Pick your herbs into individual leaves. Put some sheets of kitchen paper on trays and lay out the individual leaves separately, spreading them apart so the air can circulate. The paper helps to draw any moisture away from the leaves.

Find a place out of draughts to dry the leaves, preferably inside. Near a window is good to give them sufficient sunlight for drying – direct sunlight dries them quicker, but they fade a little.

Leave them for 1–2 days. Check on them, as they may dry sooner depending on your home: they like dry heat, no humidity. You will know when they are dry as they will feel like tissue paper and wrinkle up a little.

Once the leaves have dried, play around with mixtures to make the perfect Italian-style herb addition to your meals. I love using rosemary, thyme and oregano together in equal parts. Store them in jars or airtight containers – they will last for ages and the flavour is stronger than using fresh herbs.

DRYING IN BUNCHES

Bunches of drying woody herbs look pretty tied to a shelf or hanging above your oven. Simply gather your herbs, tie a bunch together and hang them upside down out of direct sunlight. If you want to speed up the process you may be able to dry them outside in warm, not humid, air and bring them inside when dried. If you live in a humid country, though, dry them inside in a dry place. Leave them hanging and snip off leaves as and when needed.

FREEZING IN OIL

Pick off individual leaves and place into ice-cube trays. Cover with the oil of your choice, using a ratio of one part herbs to three parts' oil. Freeze, then pop out a couple of cubes when cooking. You can use these cubes straight from frozen into a medium-hot frying pan.

Dried herbs make great
bouquet garni sachets. Try these
combinations: bay, parsley, thyme,
oregano and marjoram; or sage,
rosemary, garlic and marjoram.

No Waste Ice Cubes

Waste is everywhere and ice-cube trays make fantastic mortars for leftover fruits, herbs and more.

You can use up leftover herbs by picking them and placing them into ice-cube trays, then covering them in olive oil. Put them in the freezer, and each time you use oil for cooking, pop out a cube and use that in its place for some herb-infused oil.

For leftover fruit, simply spread individual berries, such as blueberries, raspberries or blackberries, into the trays with a sprig of mint, if you like, and top up with water. They look lovely in a jug of water or a glass of freshly squeezed juice. Fruit that's a little past its use-by date is perfect blitzed in a food processor, frozen in ice-cube trays and then used in a glass of soda water for some extra flavour.

Leftover stock or broth can be frozen in ice-cube trays and added to soups and stews – or even popped into your morning smoothies for some added protein.

Make herbs or garlic butter by mixing excess butter with chopped herbs and seasoning and placing into the ice-cube trays for use at a later date.

If you have any leftover coffee in your cafetière, pour it into an ice-cube tray and freeze. Simply pop out the cubes into a tall glass with milk for an iced coffee – perfect for a summer's day.

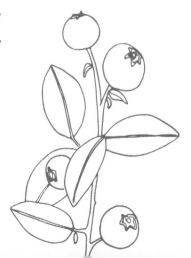

Infused Salts

Here's a fantastic way to use up the last of your ground spices and also make a great gift, presented in a well-sealed glass jar. Simply take a really good-quality salt such as a fleur de sel, pink salt or, my favourite, Murray River Gourmet salt, and your choice of ground spice. I use about 1 teaspoon of spice to about 125g of salt. Mix them together thoroughly in a small bowl, then tip into a dry airtight container. Store for as long as you would any salt and use to add a pleasing spice flavour to dishes.

SUGGESTED INFUSIONS
~ rose petals
~ lavender
~ bay leaves and dried lemon peel
~ shaved truffle
~ chopped dried chillies
~ dried orange peel and fennel seeds
~ star anise and cloves

Nut Milks and Butters

ALMOND MILK

Nut milks have grown in popularity and in turn the market has been flooded with more brands than one person has time to choose between. To add to the difficulty, more often than not these 'milks' are nothing more than water with some nuts thrown in for the sake of the name. At least by making your own you can ensure you get genuine nut milk.

MAKES 750ML–1 LITRE
260g raw almonds
950ml filtered water, plus a little
  more for pre-soaking
honey, to taste
pinch of salt

Soak the almonds overnight by placing them in a bowl and covering with about 5cm of water. The almonds will absorb water and expand. Cover the bowl with a cloth and leave overnight in a cool place. The longer the almonds soak, the creamier the almond milk, so if you prefer you can leave them to soak in the fridge for a couple of days.

When ready to prepare, drain and rinse the almonds. Place the almonds in a high-speed blender and cover with the filtered water.

Firstly, pulse a few times to loosen the almonds, then blend at the highest speed for 2 minutes. Or, if using a food processor, process for 4 minutes total, pausing to scrape down the sides halfway through. At this stage, the milk will look like fine meal and white opaque water.

Line a sieve with cheesecloth or a piece of muslin, and place over a measuring jug. Pour the almond mixture into the sieve.

Press all the almond milk from the almond meal. Gather the cloth around the almond meal and twist. Squeeze and press to extract as much almond milk as possible. Keep the almond meal for baking (for example my Veggie Scrap Crackers, page 69). You should get about 750ml–1 litre of milk. Add salt and honey to taste. Pour into a sterilised jar or bottle and keep in the fridge for up to two days.

MACADAMIA MILK

Use a super high-powered blender such as a Vitamix to blitz as the milk will split if blitzed for too long. You will always need to shake the bottle before you serve it.

MAKES 1 LITRE
800ml filtered water
160g raw macadamias
pinch of salt
1 tablespoon raw honey (optional)

Place all ingredients into a blender and blitz for no more than 30 seconds. Transfer to a sterilised jar or bottle and keep in the fridge for 1–2 days. Remember to shake well before serving.

PEANUT BUTTER

Growing up, peanut butter was the norm in our family and fights still occur to this day over whether mum should buy crunchy or smooth. This is for all of those kids who have had to endure the peanut butter wars at home. Now you can make it any way you choose.

MAKES 1 250ML JAR
300g shelled peanuts
¼ teaspoon good-quality, non-processed, non- iodised salt
1–2 teaspoons nut or extra virgin olive oil (optional)

Preheat your oven to 170°C/150°C fan/gas mark 3½. Line a tray with baking paper and roast the peanuts for 1–2 minutes or until golden brown. Allow to cool.

Tip the peanuts into a food processor. Blitz at high speed for about a minute. Scrape down the sides with a spatula, then blitz for a further 2–3 minutes or until you get the consistency you like: shorter time for crunchy, longer for smooth. Add the oil (if using) and salt at this point and blitz for a couple of seconds. Store in a sterilised jar or airtight container in the fridge for up to a month.

ALMOND BUTTER

When I was a kid, it was normal to take peanut butter sandwiches to school in your packed lunch. Often that's not possible anymore because of all of the allergies that are around the place nowadays. This is a great alternative. It's not peanut butter, but it's just as tasty and goes perfectly with jam for a nut and jelly combo.

MAKES 1 250ML JAR
300g skinned almonds
¼ teaspoon good-quality, non-processed, non-iodised salt
1–2 teaspoons nut oil (optional)

Preheat your oven to 170°C/150°C fan/gas mark 3½. Line a tray with baking paper and roast the almonds for 5–8 minutes or until golden brown. Allow to cool.

Tip the almonds into a food processor. Blitz at high speed for about 4–5 minutes. Scrape down the sides with a spatula as necessary. Add the oil (if using) and salt at this point and blitz for a couple of seconds. Store in a sterilised jar or airtight container in the fridge for up to a month.

Play around with milks and butters using extra flavourings such as food-quality salt flakes, honeys and syrups, cinnamon and spices.

Making Butter

Butter is my absolute, hands down, cannot-live-without-it favourite ingredient. All kinds of butter live in my fridge but my favourite kind is the one I churn myself, and the easiest way to do it in five minutes or less (depending on your muscle power!) is by using a jar and a marble. It also means you have control over salt content and get the byproduct of buttermilk, which you can then use in your baking.

MAKES 250G

250ml organic single cream,
  at room temperature
pinch of salt

1 litre jar and 1 large glass marble
a bowl of ice and water

Put the cream and salt into the jar and add the marble. Shake for about 3 minutes or until the cream looks softly whipped, then stiff. At first, you will be able to hear the marble rattling about in the jar. As the cream stiffens, the marble won't rattle around as much. Keep on shaking, and all of a sudden you will hear the marble again as the buttermilk starts to separate from the cream. Keep shaking until the buttermilk separates. Once it does, strain off the buttermilk. Give it another shake, remove the butter from the jar and squeeze the rest of the buttermilk out.

Once you have finished, put the butter into the bowl of ice-cold water and use your hands to massage the remaining buttermilk out – you need to remove it all or it will sour the butter. Wash the butter a couple of times, dipping it in and out of the water. Sprinkle it with some salt and mould into whatever shape you like. Put in an airtight container or wrap in clingfilm and keep in the fridge.

This is a great activity to do with kids, as they can enjoy seeing the stages from cream, to whipped cream, to buttermilk, to butter.

Carrot Top Pesto

At a farmers' market recently, I was so saddened to see person after person having their carrot tops removed and discarded. It inspired me to start digging around for ways to use them. Pesto is just one way and I can guarantee you will never throw those tops away again. Although the recipe is specific measurement-wise, the idea is to make use of what you have, so don't stress if you don't have the exact weight I've given in carrot tops. Add extra herbs, adjust the oil quantities, and play around until you get the consistency you like.

MAKES 100G

3 tablespoons pine nuts or almonds
100–150g carrot tops, washed and dried, chopped
  roughly (don't worry too much about the weight
  here, you can add extra ingredients)
small handful of any herbs (optional)
1–2 garlic cloves, minced
juice of 1 lemon (or orange) – you can add some
  zest, too, if you like
3 tablespoons Parmesan cheese, grated
olive oil
salt and freshly ground black pepper, to taste

Toast the nuts in a dry pan until they turn golden. Watch carefully and move them around the pan whilst they are cooking as they can easily burn. Put the carrot tops and herbs (if using) in a food processor and blitz until smooth. Add the garlic and lemon juice (and zest) and blitz again for a few seconds. Now add the Parmesan and enough oil to give the consistency you like for your pesto. I drizzle the oil in slowly so I can control the texture. Season to taste and blitz again for a few seconds. Store in an airtight container in the fridge for up to a week. Add more oil if too dry.

Tapenade

My Auntie Sarah had gone an entire 45 years having never eaten an olive until a recent family trip to Italy in which she had promised to try everything we put in front of her. It was quite a moment watching her face go through the motions. We all sat in anticipation waiting for a disgusted reaction. She had already made up her mind that she was going to hate it and to her surprise, quite the opposite. She said she couldn't believe she had gone so long without the salty goodness in her life, so this recipe is for her. May she never go a day without olives again.

MAKES 500G

500g good-quality pitted olives, green or black
  (I like a mix of the two)
4–6 anchovies (in oil, not brine), drained
1 garlic clove, chopped
1 small chilli, chopped
1–2 tablespoons capers, drained
3 tablespoons extra-virgin olive oil
zest and juice of ½ lemon
salt and freshly ground black pepper, to taste

Put the olives, anchovies, garlic, chilli and capers into a food processor. Blitz until smooth. With the processor still running, slowly drizzle in the olive oil. The mixture will begin to form a paste. At this point you can keep it a little chunky or continue to blitz until really smooth. Remove from the food processor and transfer into a bowl. Add the lemon zest and juice. Season to taste. Store in an airtight container in the fridge for up to a week. Add more oil if too dry.

I would certainly recommend substituting fresh tomatoes for the canned ones when in season, and making as many batches as you possibly can. To make a thicker sauce, remove the seeds prior to cooking.

Sweet and Spicy Tomato Sauce

Tomato season is always a joyous occasion for me. The smell of a freshly picked tomato always makes me salivate: tomato, a little salt – done. This recipe uses canned tomatoes for ease, but it's just the thing for using up a glut of tomatoes at the end of the season – that's when it's really time to get your sauce on. This is a very quick, easy sauce recipe. The longer you cook it, the thicker it will get. It can be used as a condiment or as a pasta sauce base.

MAKES APPROXIMATELY 500ML

80ml olive oil

4 garlic cloves, finely diced

3 x 400g cans tomatoes

3 chillies, diced (optional)

bunch of basil

2 sprigs of oregano

1–2 tablespoons brown sugar

1 tablespoon raw apple
  cider vinegar

dash of Worcestershire or chilli
  sauce (optional)

salt and freshly ground black
  pepper, to taste

Place a large frying pan over a low heat and add the oil. Add the garlic and cook until it begins to get some colour. Add the tomatoes, chillies and herbs, still on their stalks. Squash the tomatoes with a potato masher and then add the sugar and vinegar. Season to taste and cook, removing from the heat once the mixture has come to the boil. Now place a sieve over a bowl and push the sauce through, discarding the herbs.

Pour the sauce back into the pan and bring to the boil. Once boiling, reduce the heat and simmer for 20–30 minutes or until it has thickened to your liking. Pour the sauce into a sterilised jar and keep in the fridge for up to a week. You can also freeze in batches.

Pasta

Handmade fresh pasta is fantastic to make and even more fantastic to eat. Shop-bought, dried pasta just can't compare. I think that a lot of people don't try making their own fresh pasta because they think it is a really complicated and time consuming process. That is not the case with this simple recipe, which my nan taught me. When I want fresh pasta but don't want to spend forever making it, this is my go-to. You can add chopped chilli or fresh herbs to the dough to make flavoured pasta.

SERVES 2

140g plain '00' flour
140g hard wheat semolina
2 large free-range eggs
pinch of salt

Mix the flour and semolina in a bowl with your hands. Make a well in the centre and crack in the eggs, then add the salt. Mix to a dough, then turn out onto a floured surface. Knead for several minutes. If it feels too dry while kneading, add drops of water as necessary. If too wet, add a little more flour.

Knead for up to 5 minutes or until smooth. Cover with a tea towel and leave to rest for 30 minutes.

After the resting time, you can roll out the pasta. If you have a pasta machine, follow the instructions to make tagliatelle. If not, use a rolling pin to roll into sheets so thin that if you hold one up you can almost see through it. Cut these sheets into strips 2cm wide and 20cm long using a sharp knife. Cook in boiling, salted water for a few minutes or until al dente.

Be led by your hands when making pasta. Intuition will tell you when it feels right. Give it time – like all things, practice makes perfect pasta.

Bone Broth

Nourishing and absolutely delicious, bone broth is my alternative to drinking too much coffee. I aim for a cup a day to keep the doctor away. My mum said she used to drink Bonox or Oxo (gravy) powder with hot water as a kid. This is the same, but way better for you nutritionally and will help you with your salty savoury cravings. Drink it by the mug or use it as a soup base. Honestly, it is like a big, warm hug when you are feeling low. Keep some in the freezer as a backup.

Put the bones and feet into a large (5-litre) pot and add 4 litres of filtered water. Add the vinegar and let it sit in the water for 30 minutes to an hour, as the acid helps to make the nutrients more available. Roughly chop the vegetables and add to the water with the peppercorns and any other spices you may like to add – but don't add the parsley or garlic yet.

Bring the broth to the boil. Once it has reached a vigorous boil, reduce to the lowest heat and simmer until done: 48 hours for beef or pork broth; 24 hours for chicken broth. For the first half hour you need to skim scum from the top. After that, there's no need to do anything until the last half hour, which is when you add the parsley and garlic and check the seasoning.

Once cooked, drain the broth through a sieve, then add salt to taste and store in a glass jar in the fridge for up to seven days or freeze.

MAKES 3–4 LITRES

1.5kg free-range bones
  (chicken, beef or pork)
2 chicken feet or pig's trotters
  for gelatine (optional)
8 tablespoons raw apple
  cider vinegar
1 onion
2 carrots
2 celery stalks
1 tablespoon peppercorns
1 bunch parsley
2 garlic cloves
salt and freshly ground black
  pepper, to taste

*Bone broth is a wonderful base for most soups
and a perfect addition to stews and pasta sauces
for adding complexity and nutrition.*

If you don't want to use the entire piece as bacon, you can dice it into small cubes and freeze it to use as lardons in soups and casseroles.

Bacon

I love bacon almost as much as I love butter. However, I rarely feel comfortable buying bacon from the shops, as I want to know that the pig was free range. It is of huge importance to me to always buy ethically raised and humanely killed animals – there can be no exception. By buying the meat from a trusted supplier and making my own bacon I am ensuring traceability that often does not come with shop-bought, pre-packaged bacon.

Use a pestle and mortar to crush all the spices and tip into a bowl with the remaining ingredients, except the pork. Mix together. Rub the mix over the pork belly, top and bottom. Place in a ziplock bag and squeeze out the air as you lock it.

Place in the fridge for seven days and turn the bag over daily. Liquid will be drawn from the pork: this is good. After the seventh day, remove from the bag and wash the pork with cool water. Pat dry with kitchen paper and sit it on a wire rack over a tray in the fridge, uncovered, for a further 24 hours.

Once done, preheat your oven to its lowest setting, and cook for 90 minutes. Remove and allow to cool. Keep in the fridge in an airtight container and slice as you need it. It will keep for three weeks in the fridge or freeze for up to six months.

MAKES 800G–1KG

1 tablespoon black peppercorns
2 teaspoons juniper berries
1 teaspoon chilli flakes
3 tablespoons good-quality, non-processed, non-iodised salt
3 garlic cloves
60ml honey
2 teaspoons rosemary
1 teaspoon thyme
3 bay leaves
2 tablespoons cold, strong coffee
1kg pork belly

Potted Pork

Perfect for your cheeseboard or sandwich, this is melt-in-your-mouth goodness. The same recipe can be used for fish or slow-cooked meat, duck or even chicken. You just need to slow cook the meat or fish so that it shreds, then use the same method from the remaining recipe.

SERVES 6

butter, for greasing

400g free-range boned pork (neck and shoulder are good cuts)

splash of olive oil

1 heaped tablespoon salted capers

2 tablespoons finely chopped fennel fronds

pinch each of salt and freshly ground white pepper

320g unsalted butter, clarified and cooled

salt and freshly ground black pepper, to taste

TO SERVE

½ cucumber

splash of raw apple cider vinegar

Preheat the oven to 150°C/130°C fan/gas mark 2. Butter six ramekins or small dishes.

Season the pork with oil and salt, and cook in the oven for a couple of hours or until the meat is tender and falling apart. Remove from the oven and pull the pork into shreds. Add the capers, fennel, white pepper and a little salt to the meat. Mash it all to a rough paste with a fork.

Divide the meat among the prepared ramekins or dishes, press down to level the surface and cover with two-thirds of the clarified butter. Transfer the ramekins to a shallow ovenproof dish and add boiling water to come halfway up the sides of the ramekins. Cook in the preheated oven for 10 minutes. Remove from the oven and leave to cool.

Pour the remaining clarified butter over the meat until it is completely covered. Refrigerate the ramekins for at least 3 hours to set properly.

Remove the potted pork from the fridge 20 minutes before serving to bring it back to room temperature. Serve straight from the pots or turn the pork out onto a plate so that the butter layer is underneath.

Just before serving, slice the cucumber diagonally, splash with apple cider vinegar and sprinkle with salt. Serve the potted pork with the cucumber slices.

HOW TO CLARIFY BUTTER

In a small saucepan, heat the butter over a low heat. Ensure it does not brown or indeed burn. The butter will separate into white milk solids and liquid golden butter. The milk solids will sink to the bottom of the pan, then you just need to pour the liquid butter into a jug or container and discard the solids.

Cured Fish

Cured fish reminds me of my two very dear Swedish mates Beata and Carolina. I have to be honest with you, it's not my most favourite of things, but when I make it for others, they love it. This recipe makes me think of my friends on the other side of the world and the kilos of gravlax they would eat on a weekly basis. It is also a fantastic way of adding some shelf life to fish you may have left over.

SERVES 20 AS AN ENTRÉE OR
40 AS HORS D'OEUVRES

2kg sustainably sourced
  fish fillets, such as salmon,
  monkfish, snapper, cod or
  mackerel, unskinned
220g sugar
115g salt
1 tablespoon freshly ground
  pepper (I like to use white or
  pink pepper)

Rinse the fish and pat dry with kitchen paper. Cut each fillet horizontally into two equal halves.

Combine the sugar and salt, then cover both sides of each fillet half with the mixture. Sprinkle the flesh side of each piece with ground pepper. Place each fillet half, flesh side up, in a dish just large enough to hold all the fish.

Cover the dish lightly with clingfilm and leave the fish to marinate at room temperature until the sugar-salt mixture has melted into the fillets. (This should be no longer than 6 hours. Skip this step entirely if you are curing your fish in hot weather.)

Place a small pan or plate on top of the clingfilm-covered fish. Weight the plate lightly, using some unopened tins of food. Refrigerate the weighted gravlax for at least 48 hours and up to a week. Every 12 hours, turn the fish over in the brining liquid that has accumulated in the bottom of the dish to ensure that all parts are evenly marinated. Re-cover with clingfilm and return the weighted pan to the fridge.

When you have finished marinating the fish, remove it from the fridge. Using a sharp knife, cut the cured fish into paper-thin slices, pulling each slice away from the skin. The fish can be stored in the fridge for up to a week and in the freezer for up to a month.

green Curry Paste

Green curry was my mum's idea of an adventurous dinner when I was a kid. I remember the first time we ever had it so clearly – we were usually a meat-and-three-veg household. It was very obviously from a jar back then, but these days, as a keen cook, my mum would definitely use this recipe. Making your own gives you the freedom to tweak the recipe to suit what you do and don't like.

Put everything except the coconut cream into the food processor and blitz. Keeping the processor running, add the coconut cream a tablespoon at a time until you have added enough to make a smooth paste. Store in an airtight container in the fridge until ready to use.

Use in recipes in place of shop-bought curry paste.

This will keep for a few months in the fridge.

MAKES ABOUT 80ML
(enough for two curries, depending on how spicy you like it)

2 shallots, chopped
4 lemongrass stalks, minced
4 green chillies, deseeded
  and chopped
10 garlic cloves
4cm piece of fresh ginger, sliced
1 teaspoon ground cumin
1 teaspoon ground coriander
small handful basil, chopped
small handful coriander, chopped
1 tablespoon fish sauce
2 tablespoons soy sauce
zest of 1 lime and juice of half
4 tablespoons coconut cream

*This paste can be made in larger batches and frozen
if you prefer to make more at a time.*

Red Curry Paste

I actually learnt how to make a few different kinds of curries on a trip to Sri Lanka. It was magical being in a place where so many incredible spices grow, and to play around with them in a giant pestle and mortar was the highlight of my trip. Whilst this recipe is not a traditional Sri Lankan curry paste, the same principals apply in making the paste. By making it yourself you can control the heat and spice and take out things you don't love. You can always double or triple the batch and store it in the fridge for regular use.

Put everything except the coconut cream into the food processor and blitz. Keeping the processor running, add the coconut cream a tablespoon at a time until you have added enough to make a smooth paste. Store in an airtight container in the fridge until ready to use.

Use in recipes in place of shop-bought curry paste.

This will keep for a few months in the fridge and can be made in larger batches and frozen if you prefer to make more at a time.

MAKES 250G
(enough for two curries, depending on how spicy you like it)
2 shallots, chopped
2 lemongrass stalks, minced
2 red chillies, chopped
  and deseeded
8 garlic cloves
4cm piece of fresh ginger, sliced
4 tablespoons tomato purée
2 teaspoons ground cumin
1 teaspoon ground coriander
½ teaspoon white pepper
4 tablespoons fish sauce
  or soy sauce
2 teaspoons shrimp paste
2 tablespoons sugar
3 tablespoons chilli powder (less
  if you prefer a milder paste)
5 tablespoons lime juice
4–6 tablespoons coconut cream

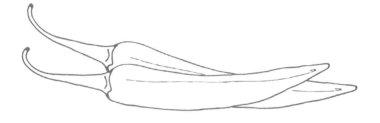

Infusing Oils and Vinegars

INFUSED OILS

Herb-infused oils are not just great for cooking, they also make lovely gifts and some can be used as massage oil or added to the bath. You can use any oil you like for this – it's best to use one you like to cook with (or want to smell like!).

Sterilise and thoroughly dry a large jar or bottle. Fill it with herbs and chillies, garlic or edible flowers, depending on what you want to use the oil for.

Slowly pour oil over the herbs. Use a skewer or similar to move the herbs around to ensure no air pockets remain. Add enough oil to completely cover all the herbs, filling your vessel right up to the brim.

Tightly seal the jar or bottle, give it a few shakes, and put it in a cool, dark place. Shake every few days and let it sit for at least 4–6 weeks so the herbs can infuse. After that time, strain the oil into storage bottles through a cloth-lined sieve. Give the herbs a squish before discarding.

Seal and label your bottles. Store in a cool, dark place for up to two years.

INFUSED VINEGARS

Infused vinegars are wonderful to cook with and make fantastic medicinal tinctures. Feel free to experiment with different herbs and vinegar types.

Sterilise and thoroughly dry a large jar or bottle. Fill it with herbs and chillies, garlic, fruit or edible flowers, depending on what you want to use the vinegar for.

Heat the vinegar over a low heat until simmering. Slowly pour the hot vinegar over the herbs. Use a skewer or similar to move the herbs around to ensure no air pockets remain. Add enough vinegar to completely cover all the herbs, filling your vessel right up to the brim.

Tightly seal the jar or bottle, give it a few shakes, and put it in a cool, dark place. Shake every few days and let it sit for at least 4–6 weeks so the herbs can infuse. After that time, strain the vinegar into storage bottles through a cloth-lined sieve. Give the herbs a squish before discarding.

Seal and label your bottles. Store in a cool, dark place for up to two years.

These make the most thoughtful and pretty gifts for any food lover. Choose bottles in a variety of shapes and sizes for a decorative addition to your kitchen.

*Once you get the hang of these crackers you can
experiment with all kinds of spices, nuts and seeds,
and even make sweet versions for a healthy snack.*

Veggie Scrap Crackers

My incredibly talented cousin, Yasmin, who is one of the greatest chefs I have ever worked with, inspires this wonderful recipe. She does not waste anything and that is where these crackers came into being. If you don't own a juicer, you can just grate up some veggies or use leftover ends of things – just make sure there is not too much moisture in them. This recipe does work best in a dehydrator, but you can use an oven on low heat for almost the same result – it's just a longer process.

MAKES 8–10 CRACKERS

400g golden flaxseed
200g almond meal
125g shelled pumpkin seeds
125g veggies, grated, or
  juicer scraps
1 teaspoon sea salt

If you're using the oven, preheat it to its lowest setting and line a large baking tray with baking paper. If you're using a dehydrator, set it up to 46°C and line the tray.

Place the flaxseed in a bowl and just cover with filtered water. Soak for one hour.

After soaking, combine with all the other ingredients in a large bowl and form into a big ball using your hands. If the mixture seems too dry, add a little water, or if it is too sticky, add a little more almond meal. Place on the lined tray and begin flattening the dough out with a rolling pin (or just use your hands) until the dough is about 2mm thick.

If you're using the oven, transfer the tray to the oven and bake overnight until firm. Cool on the tray and then break into pieces. If you prefer straight and precise pieces, cut squares into the dough before baking.

If using the dehydrator, cook overnight, then take off the paper in the dehydrator tray the next day and cook on the opposite side for another two hours. You only need to do this if the cracker is not solid.

These will keep for about one week in an airtight container, but like all crackers are best if consumed within a couple of days.

Fermenting

SAUERKRAUT

My cousin, Sam, taught me everything I know about fermenting. He has a wonderful fermentation business called Gut Feeling and together we have been running fermentation workshops all over Australia. It has been really magical for us to pass on such an old art that creates a food so good for us and that's easy to master at home. For best results use local, organic or unsprayed ingredients in all of your ferments. When washing cabbages and other produce for fermenting I also recommend using filtered water.

MAKES 500G

500g green or red cabbage (keep one intact leaf to act as a bung for the top of the jar)

10g sea salt (good quality, not iodised – I use Murray River salt)

100g root vegetables and/or green apple, grated

pinch or two of herbs or spices (juniper is traditional, and seeds such as caraway, dill, coriander and mustard work well)

1 500g sterilised jar

Slice the cabbage, cutting against the grain to create thin, medium-length strips, and place in a large bowl. Add your sea salt. With clean hands (but not having used hand sanitiser, as this will kill good bacteria), start to massage the salt into the cabbage for 3–4 minutes. Set aside for the salt to pull the juices from the cabbage. Prepare by julienning any other vegetables into similar-sized strips, and by getting the herbs and spices you like ready to be added to the cabbage.

After 5–10 minutes the salted cabbage will have released more juices and wilted. Continue to massage the cabbage until it has softened and a reasonable amount of brine has been created. Add your other vegetables, herbs and spices to the bowl and mix thoroughly.

Now pack into your sterilised jar, using your hands to press the vegetables down tightly. Ideally you want them packed to within 2–3cm from the top of the jar, with brine covering all the vegetables.

Fold a small portion of the reserved cabbage leaf and place on top of the vegetables to keep them submerged in the brine, but ensure there is still room to put a lid on. Wipe any bits from the rim of the jar using paper towel and put on the lid. Leave to ferment at room temperature for at least four days and up to three weeks. For a ferment using this technique I find 7–10 days works well.

Remember that gas (carbon dioxide) is released during the fermentation process, so it is important to open the lid to 'burp' the jar every day or two to release the gases. Just be quick to reseal it so oxygen doesn't get a chance to sneak in and spoil your ferment. So you will literally open the lid but not take it off or lift it up. I suggest burping on days three, five and seven. After this, remove the cabbage leaf and taste the kraut. If you are happy, replace the lid, refrigerate and enjoy within two months.

A FEW THINGS ABOUT SAUERKRAUT

~ Sauerkraut, which in German means sour cabbage, is an age-old method of preserving cabbage and other ingredients by fermentation. 'Kraut' was traditionally prepared to provide vital nutrition in northern Europe during the cold winters when vegetables could not be grown. The same technique is still used today with more knowledge of the diverse health benefits associated with fermented foods.

~ Captain Cook is said to have conquered scurvy when he took several barrels of sauerkraut on board his ocean-going ships, and subsequently didn't lose any of his crew members to the disease. This is due to the high vitamin C content in sauerkraut.

~ More than 2,000 years ago, Chinese labourers who were building the Great Wall of China supplemented their rice diet with servings of cabbage preserved in wine. This later ended up in Europe when Genghis Khan took it along with them and introduced it. This is said to perhaps be where Europeans found their love for kraut. Most people could not afford the luxury of wine so made their own version of sour cabbage using salt.

~ Unpasteurised or 'live' sauerkraut is considered a superfood. It contains large amounts of vitamins C, B and K, as well as vitamin A, folic acid and important minerals such as calcium, potassium, iron, phosphorus, sodium and magnesium. Live sauerkraut also contains various lactic acid strains of bacteria, which can help to restore a healthy population of beneficial bacteria in the digestive tract. This is essential because recent research links an unhealthy gut to acne, depression and obesity, among other unwelcome conditions, whereas a more balanced 'gut flora' can lead to enhanced immune system function, a less inflamed digestive system and the ability to fight off disease. Eating sauerkraut is a fantastic way to remain well.

CARROT AND GINGER SAEURKRAUT

This is my favourite combination. Use the recipe on page 70 as your basis. To appropriately salted, prepared cabbage, add a handful of grated carrot, 1 teaspoon finely grated or chopped root ginger and ½ teaspoon finely grated or chopped garlic.

TIPS FOR A HAPPY FERMENT
~ Use good-quality, fresh ingredients.
~ Salt veggies properly using 2–3 per cent good-quality sea salt per total weight of vegetables.
~ Use thoroughly clean hands and equipment.
~ Keep oxygen out of your jars.
~ Ferment for at least four days, but longer is better.
~ Feel empowered to experiment with ingredients, although about 70 per cent cabbage ensures a consistent result.

KIMCHI

Kimchi is Korea's fiery version of fermented vegetables. As with sauerkraut, for best results use local, organic or unsprayed ingredients to make kimchi and wash everything in fermented water if possible.

MAKES 500G

500g green or red cabbage, sliced (keep one intact leaf to act as a bung for the top of the jar)
12g sea salt (good quality, not iodised – I use Murray River Salt)
½ small handful of grated carrot
½ small handful of grated daikon radish
1 onion, either spring or red, finely diced
½ teaspoon finely grated or chopped garlic
½ teaspoon finely grated or chopped ginger
½ teaspoon finely grated or chopped red chilli

1 500g sterilised jar

Prepare the cabbage and salt using the method for making sauerkraut. Mix all the other ingredients in a small bowl. Add to your brined cabbage and pack into a clean jar as described on page 70.

Gut health is instrumental to good well-being and it is so easy and economical to make at home. All you need for a good sauerkraut is cabbage and salt. The rest is just a little therapeutic massage and patience.

Chutneys

FIG AND PEACH CHUTNEY

Summer in a jar. The two are seriously best friends. You can swap out the peaches for plums or nectarines too. If you can only get your hands on ripe figs that is also fine, just throw it all in and cook until you get the consistency you love.

MAKES 1–2 LITRES

500g unripe figs
500g peaches
500ml raw apple cider vinegar
500ml malt vinegar
250g soft brown sugar
500g brown onions
125g crystallised ginger
175g honey
2 tablespoons salt
1 tablespoon cardamon seeds, bruised
1 teaspoon freshly ground white pepper

4–8 500g sterilised jars

Wash the figs and pat dry. Cut off and discard their tops and cut the fruit into quarters. Wash, stone and quarter the peaches.

Combine the figs, peaches, vinegars and sugar in a large, heavy-based saucepan and stir until the sugar is dissolved. Add the remaining ingredients and bring to the boil, then simmer for 1–2 hours, or until the fruit is tender and falling apart.

Pour into sterilised jars, seal and allow to cool. Label and store in a cool, dark place for up to three years. For the best flavour, let the chutney mature for at least three months before eating it. Refrigerate after opening.

BEETROOT CHUTNEY

Perfect on a cheeseboard or in a sandwich, this is an everything chutney. I love beetroot and this chutney is a wonderful way to use up the summer glut with a little more flavour than pickled beets. If you prefer fewer spices you can actually take out the ones you don't like and keep it simple.

MAKES 1–1.5 LITRES

750g beetroot
olive oil
3 teaspoons fennel seeds
3 teaspoons yellow mustard seeds
1 brown onion, finely diced
zest and juice of 1 orange
500ml balsamic vinegar (or use raw apple
  cider vinegar for lighter flavour)
250ml water
500g soft brown sugar
2 cloves
5cm strip orange rind
salt and freshly ground black pepper to taste

2–3 500g sterilised jars

Peel the beetroot and chop into small chunks. Heat the tiniest amount of olive oil in a frying pan and fry the fennel and mustard seeds for about 30 seconds, just until the aromas are released.

Add all of the ingredients to a large, heavy-based saucepan over a medium heat, and bring to the boil. Cook for 30 minutes or until the beetroot is soft and the liquid has reduced and thickened. Pour into sterilised jars, seal and allow to cool. Label and store in a dry, cool place for up to a year. Refrigerate after opening.

PEAR, PLUM AND PRESERVED LEMON CHUTNEY

This is for the gluts of pears, stone fruits and lemons. You can use any variety of all of these fruits, which makes it the perfect neighbourhood swap chutney. Ask your neighbours what they have and swap accordingly – then maybe you can even swap jars afterwards for some fun.

MAKES 1.5–2 LITRES

250g red onions, diced

200ml red wine vinegar

150ml raw apple cider vinegar

250g granulated sugar (or use brown sugar for a richer flavour)

1kg plums

500g pears

80g preserved lemons (see page 82), rind only, sliced finely

1 teaspoon pink or white peppercorns, finely ground

1 teaspoon cinnamon

½ teaspoon nutmeg

½ teaspoon mixed spice

1 teaspoon salt

4 500g sterilised jars

Put the onions in a large, wide-based saucepan over a high heat with the vinegars and sugar. Once boiling, lower the heat and gently simmer, stirring occasionally, for about 10 minutes until the onions soften and become shiny.

Wash, stone and dice the plums. Wash, core and chop the pears into small pieces. Add the fruit to the pan along with the preserved lemons and the spices and salt. Bring back to the boil, then lower the heat to a simmer and stir occasionally so that it doesn't stick to the bottom and burn. Cover the pan and continue to cook over a low heat for an hour until the mixture becomes thick and has reduced by a third.

Pour into sterilised jars, seal and let cool. Label and store in a cool, dark place for up to a year. Refrigerate after opening.

You can play around with the aromatics and use whatever tickles your fancy.

APPLE, PEAR AND TOMATO CHUTNEY

This chutney is so easy to make and you can just portion the recipe
if it is too much. However, if you make this large batch you can
easily jar it all up and keep in the pantry for those last-minute
visitors or birthday gifts. Feel free to adjust and use seasonal produce
or different spices. It really is an anything-goes chutney.

MAKES 3–4 LITRES
(you can halve this recipe
if you would prefer to make
a smaller amount)

1kg brown onions
1kg red onions
1kg apples, cored
1kg pears, cored
4.5kg tomatoes (preferably
  Roma or vine)
750g raisins or sultanas
125g fine white salt
770g white sugar
250g soft brown sugar
750ml raw apple cider vinegar
6 garlic cloves, crushed
80g fresh ginger, crushed
2 teaspoons fennel seeds
2 teaspoons ground cinnamon
large pinch of chilli flakes
8 cardamom pods, bruised
zest of 2 lemons

16–20 250ml sterilised jars

Dice the onions and unpeeled apples and pears very finely and the
tomatoes a little chunkier (you can do all of this in a food processor if
you like). Cook them in a large, heavy-based saucepan over a medium
heat for about 10 minutes – there's no need to add any water, as it's
already quite runny.

Add the remaining ingredients and simmer gently for 2–3 hours.

Drain any excess liquid into a smaller saucepan, reduce it over a high
temperature until thickened and syrupy, return it to the mixture and heat
again. (If you prefer you can actually just bottle this liquid for use in
sauces.) It can be quite a liquid chutney, depending on the tomatoes'
water content.

Pour into sterilised jars, seal and allow to cool. Seal, label and store in a
cool, dark place for up to three years. For the best flavour, let the chutney
mature for at least three months before eating it. Refrigerate after opening.

Pickles

BUTTERMILK PICKLES

What a wonderful pickle these are. They have such a gorgeous flavour, different from those made with vinegar. You really can use almost anything for this recipe. I love to do just pears in a buttermilk pickle with some bay leaves and cinnamon sticks in the jar.

MAKES 2–3 LITRES

2kg root vegetables and cabbages of your choice, roughly chopped, or use a glut of pears, quartered
1 litre buttermilk (if you have made the butter on page 44, use the drained buttermilk)
salt
water

3-litre sterilised jar

Put the chopped vegetables or pears into a large, clean glass jar or a ceramic container that can be sealed. Pour over the buttermilk. Mix the vegetables into this liquid and top with a scrunched up cabbage leaf so that they are compressed and just covered with the liquid. Seal the jar and leave to ferment in a warm place for 12 hours.

Uncover and add salt water if the vegetables are not covered in liquid (allow 3 teaspoons of salt for every litre of water). Put the weight on top of the vegetables again and seal the jar.

Return the jar to the same warm place to ferment for three days. After this time you can bottle the vegetables and liquid as you like and refrigerate. Unlike many foods, these benefit from keeping. Enjoy as a side dish with almost anything.

PICKLED COURGETTES

This pickle recipe can also be used for other delicate vegetables such as cauliflower and cabbage. Always remove the pulp in the middle of the courgettes or you will end up with mush in a jar.

MAKES 1.5 LITRES

1.2kg different-coloured courgettes
2 teaspoons yellow mustard seeds
1 teaspoon black mustard seeds
1 teaspoon ground turmeric
1 teaspoon fennel seeds
220g white sugar
500ml raw apple cider vinegar
250ml white vinegar
½ teaspoon chilli flakes

4–6 250ml sterilised jars

Wash and dry the courgettes and slice lengthways. Using a spoon, remove and discard the soft seeds in the middle. Slice the courgettes finely and put in a large heatproof bowl.

Combine the remaining ingredients in a large saucepan and bring to the boil. Boil for about 20 minutes.

Pour this boiling liquid over the courgettes and leave for about 5 minutes, then transfer the lot back into the saucepan and cook for a further 3 minutes.

Ladle into sterilised jars, seal and let cool. Label and store in a cool, dark place for up to two years. For the best flavour, let the pickle mature for at least three months before eating it.

PICKLED CHERRIES

This is my favourite pickle recipe ever. I love to use the pickles in sweets and savouries, in particular my chocolate cherry pie. They are a wonderful addition to a cheeseboard or cocktails, or even with meats as a condiment. You can really taste the difference regionally when you use cherries from different places too.

MAKES 1–1.5 LITRES

1kg cherries

6 strips of zest and the juice of 1 orange

350g caster sugar or brown sugar or mixture

500ml white wine vinegar or raw apple cider vinegar

180ml water

10 cloves

2cm piece of fresh ginger, smashed

2 cinnamon sticks

6 cardamon pods, bruised

4–6 250ml sterilised jars

Wash the cherries and remove any stems and the stones. Combine all the ingredients except the cherries in a saucepan, bring to the boil, then simmer for 20 minutes to allow the flavours to infuse. Add the cherries and cook for 2 minutes, or until tender. Alternatively, for a firmer fruit, skip cooking the cherries and pack them straight into sterilised jars and pour the hot liquid over the top. This is completely up to you.

Spoon the cherries into sterilised jars and add enough hot pickling liquid to cover them. Seal and allow to cool. Label and store in a cool, dark place until ready for use. These will store for up to two years. Refrigerate after opening.

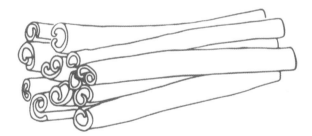

SPICY PICKLED ONIONS

These don't really need an introduction! You either love them or hate them. I sit firmly in the 'love' camp, as does my entire family, who always fight over the few jars my nan brings up at Christmas time.

MAKES 1.5KG

1.25kg pickling onions
60g fine white salt
1 litre water
2 litres white wine vinegar
2 litres brown malt vinegar
  (or use half malt and half raw apple cider vinegar)
65g firmly packed brown sugar
4 garlic cloves, smashed
2cm piece of fresh ginger, bruised
1 cinnamon stick
3–5 dried hot chillies

FOR THE JARS

yellow mustard seeds
brown mustard seeds
fresh or dried bay leaves
cinnamon sticks

4-6 250g sterilised jars

Place the onions in a large bowl and cover them with boiling water to loosen their skins. Leave until the water is cool, then drain and peel. Place them in another large bowl and add the salt and 1 litre water. Cover with a tea towel and leave for 24 hours. The next day, wash and dry the onions and set aside.

Combine the vinegars, sugar, garlic, ginger and cinnamon stick. Bring to the boil, then boil for 10 minutes.

Spoon the onions into sterilised jars, then add some of the decorative ingredients such as mustard seeds, bay leaves and cinnamon sticks. Strain the vinegar mixture into a jug, then pour into the jars to cover the onions. Top up the jars with more decorative spices, then seal and let cool. Store in a cool, dark place. For the best flavour, let the onions mature for at least three months before eating. Refrigerate after opening.

Preserved Lemons

When life gives you lemons, my new motto is: preserve them! Use them anywhere you want some bitter and acidic addition to your cooking. They are great mixed through salads, pushed under the skin of a chook before roasting or stirred into your boiled rice. Just give them a quick rinse and pat dry before using.

Keep a few lemons aside for extra juice. Quarter the other lemons, keeping their bases on. Squeeze some of the juice (discarding pips) from each lemon into the jar so you have some juice and set aside. Rub a generous amount of sea salt into the centre of each lemon (at least a teaspoon).

Pack the prepared lemons into the sterilised jars with the bay leaves, a cinnamon stick and a teaspoon of fennel seeds in each jar. Sprinkle another 2 tablespoons of sea salt into each jar. Add enough lemon juice to cover the lemons.

Seal and store in a cool place. Try to store for two months for maximum flavour and preservation.

When ready to use, just rinse with water then cut away and discard the flesh and slice into the sizes you need. They are great in slow-cooked dishes and sliced super thin in salads.

MAKES 4 250ML JARS
12–14 lemons
220g sea salt
10–20 fresh bay leaves
4–6 cinnamon sticks
4 teaspoons fennel seeds

4 x 250ml sterilised jars

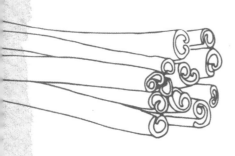

Jams and Curds

ANY PLUM JAM

I love to add different flavourings to this jam such as orange blossom water or culinary lavender.

MAKES 1 LITRE

1.5kg plums (include some under-ripe ones)
1 teaspoon cinnamon
1 teaspoon mixed spice
300ml orange juice
1.25kg sugar (brown gives more depth)

4–6 250ml sterilised jars

Wash, rinse and halve the plums. Remove the stones and set aside – you will need them for their pectin.

Put all of the ingredients except the sugar and plum stones into a wide-based saucepan. Put the stones into a square of muslin cloth, tie with kitchen string, then add to the pan too. When ready to cook, place a small saucer in the freezer (for testing the jam later).

Bring the mixture to the boil, then simmer over a medium-low heat for about 20 minutes until the skins fall off the plums. Reduce the heat to low. Add the sugar and stir rapidly until it dissolves. Return to a rolling boil and cook for a further 10–15 minutes or until it has reached setting point (104.5°C on a sugar thermometer). To test the jam, put a tiny dollop onto the frozen saucer and leave for a minute. If the jam is set, it will form a 'crinkle' on the surface when you touch it. If not, cook for a few more minutes and try again. Once cooked, remove from the heat. Discard the muslin bag and its contents. Pour the jam into hot sterilised jars, seal and let cool. Label and store in a cool dark place for up to two years. Refrigerate after opening.

ANY BERRY JAM

Mix the berries up or keep them to a single flavour. Best to put in some under-ripe ones as berries are the lowest in pectin of the fruits so a good set may sometimes take longer. Try adding a small handful of chopped fresh mint or a tablespoon of rosewater to the jam.

MAKES 1 LITRE

1kg berries of your choice
900g granulated sugar
juice and zest of 1 lemon

4–6 250ml sterilised jars

Wash, hull and pat dry the berries. Cut the strawberries into quarters and leave any other berries whole. Put them into your jam pan (a wide-based saucepan is fine). Add the sugar, lemon juice and zest (this helps the set). Stir well and, if you have time, let this sit for a few hours lightly covered with a tea towel.

When ready to cook, place a small saucer in the freezer (for testing the jam later). Now put the pot onto a very low heat, stirring occasionally until all the sugar is completely dissolved. Turn up the heat until the jam is at a rolling boil and begins to froth.

Cook for about 20 minutes or until set, stirring every couple of minutes or so to prevent burning. To test the jam, put a tiny dollop onto the frozen saucer and leave for a minute. If the jam is set, it will form a 'crinkle' on the surface when you touch it. If not set, cook for a few more minutes and try again. Once cooked, cool for a few minutes and then put into hot, sterilised jars. Label and store in a cool, dark place for up to two years.

CITRUS CURD

Good on toast, in tarts, over ice cream, stirred through yogurt or straight from the jar. This is my mum's version of my recipe and she wins. It's excellent.

MAKES 750G–1KG
375g unsalted butter, cubed
zest and juice of 6 lemons, or a mix
  of lemons and limes
500g granulated sugar
25g cornflour
6 free-range egg yolks

4–6 250ml sterilised jars

Put the butter in a saucepan over a low heat and add the zest, juice and sugar. Stir until the sugar is dissolved, then bring to the boil. Mix the cornflour with 50ml water in a small cup, then add to the lemon-butter mixture. Bring back to the boil, stirring constantly for a couple of minutes.

Mix another 50ml of water with the egg yolks, remove the lemon mixture from the heat and stir in the egg yolk mixture. Push through a sieve into a sterilised jug and pour into your sterilised jars, leaving 0.5cm head space and pushing out any air bubbles with a jam spatula or skewer. Label and store – this will keep in a cool, dark place for up to six months.

FRUIT BUTTER

Really this recipe came from a mistake. I was trying to make rhubarb jam and it came out more like a butter or curd by way of its consistency, but in the end I liked it more. You don't have to add the floral water and can add a pinch of cinnamon or mixed spice if you prefer.

MAKES 1 LITRE
1kg rhubarb or apples
440g granulated sugar
zest of 1 orange
85ml orange juice
15ml orange blossom water or rose water

4–6 250ml sterilised jars

Wash the rhubarb and discard the ends or core the apples and chop the fruit into small pieces. Combine all ingredients into a wide, heavy-based saucepan with 125ml water. Cook over a medium–low heat for about 45 minutes, stirring occasionally to stop it sticking, until you have a dark and thickish paste. Pour into sterilised jars, label and store in a cool, dark place for up to six months.

Lemon curd can be frozen in your jars too by using freezer lids. Leave a good 1cm head space to allow for expansion.

Vanilla Ice Cream

This is the purest form of ice cream and the joy of it is that you don't need an ice-cream maker, nor do you need to churn it in any way. Just make and freeze.

SERVES 1–2

3 tablespoons runny honey
3 free-range egg yolks
250ml double cream
seeds scraped from ½ vanilla pod
  or 1 teaspoon vanilla extract

Heat the honey in a saucepan till just warmed. Put the egg yolks in a medium-sized bowl and whisk in the warm honey. Whip in the cream and the vanilla. Pour into a freezer-proof dish. Freeze for 2–3 hours or until firm. Remove from the fridge 5–10 minutes before serving.

Try using different types of honeys or maple syrup
for more variety in your ice cream.

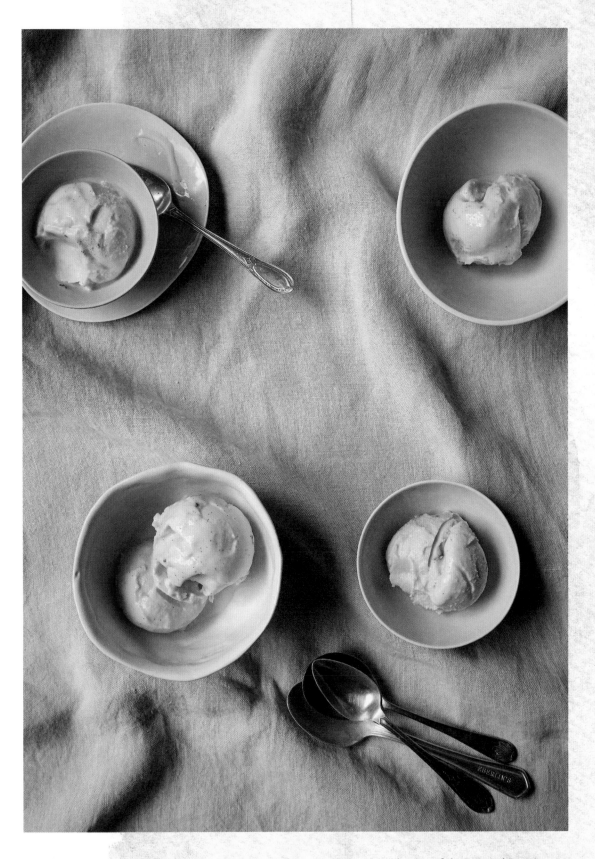

*Try these sorbets scooped into small balls and added
to champagne flutes, before topping with your favourite sparkling
wine, Prosecco, Cava or Champagne.*

Pure Fruit Sorbets

Along with making chutneys and preserves, this is a great way to use up gluts of seasonal fruit. Just peel or hull, if needed, chop into small pieces, freeze and then purée in a food processor or by using a hand blender. Add a little icing sugar or honey if you wish to sweeten the fruit to your liking, and give it your own twist by adding spices, herbs or a hint of booze or juice. These are some of my favourite combinations:

MANGO, LIME AND HONEY

Blend chunks of frozen mango with a hint of lime juice and a little honey until smooth. Freeze for 10–20 minutes once puréed for a firmer texture.

SPICED PEAR AND BLACKBERRY

Blend frozen chunks of peeled and cored pears with a handful of blackberries, a drop of maple syrup and a pinch of ground mixed spice until smooth. Freeze for 10–20 minutes once whipped up for a firmer texture.

SUPER STRAWBERRY

Blend hulled, frozen strawberries with a few teaspoons of icing sugar, to taste, until smooth. The strawberries will need a little extra blending as their texture isn't as smooth and creamy as mangos or pears, but keep blending and it'll get there. Freeze for 10–20 minutes once puréed for a firmer texture.

Natural Food Colouring

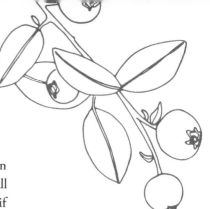

My brother, Chris, was the poster child for what goes wrong when kids eat artificial colours. Red cordial was his theatre. This is for all of the mums out there who have to put their children on leashes if they come into contact with party food.

YELLOW

120ml water
½ teaspoon grated fresh turmeric (or use powdered)

Boil the water and turmeric in a small saucepan for 5 minutes. Allow to cool fully and store in a jar or airtight container in the fridge for up to a month. Use a teaspoon at a time – more for an intense yellow.

PURPLE

70g blueberries, fresh or frozen (if frozen, thaw and let them dry out on kitchen paper)
4 teaspoons water

Blitz the berries and water in a food processor until smooth. Strain through a super-fine mesh sieve to separate the solids from the colour. Store in a jar or airtight container in the fridge for up to a month. Use a teaspoon at a time – more for a brighter purple.

GREEN

60g spinach
6 tablespoons water

Cover the spinach in water and boil for 5 minutes. Strain and discard the cooking liquid. Blitz the spinach and 6 tablespoons water in a food processor until smooth. Add more water if it does not combine properly, a dash at a time. Strain through a fine sieve, allow to cool fully and store in a jar or airtight container in the fridge for up to a month. Use a teaspoon at a time – more for a darker green.

RED

80g raspberries
4 teaspoons water

Blitz the berries and water in a food processor until smooth. Strain to remove seeds. Store in an airtight container for up to two weeks in the refrigerator. Start by adding a teaspoon to icings, frostings or batter to give a subtle red hue. Add more colouring gradually.

PINK

125g cooked beetroot (sold in vacuum packs)
2 teaspoons juice from the pack

Blitz the beetroot and juice in a food processor until smooth. Strain if desired. Store in an airtight container for up to two weeks in the fridge. Add 1 teaspoon to icings, frostings, or batter for starters to impart a pink hue. Add more colouring gradually, if needed.

Drinks Trolley

LOOK, I'LL BE HONEST, I love a good glass of wine or two. But there is more to a drinks trolley than just beer or Riesling. Every time I go to a bar (which is not as often as I would like these days), there seems to be a new cocktail on the list with a fancy new ingredient I have never heard of. But often these unheard-of ingredients are actually things that have been around for decades, even centuries, if not more, just reinvented with a fancy-sounding name (and price tag to match).

Cocktails are fun and delicious, but often filled with sugar, leaving you with that headache the next morning, which is only explainable as the sugar hangover. The older I get, the less I can drink and sometimes I want to feel social but don't want to drink wine. So my new best friend is homemade kombucha or ginger beer (in a wine glass, of course). It gives me that same little kick of energy as wine does but rather than wrecking my gut I am actually helping it. With these, and the other recipes in this chapter, like the cordials and shrubs, lemonades and citroncellos, you can control which ingredients and how much sugar goes into them. No more sugar headaches for me or you!

Kombucha

Kombucha is the name of a naturally cultured, therefore slightly fizzy, tea that has been made and drunk in different parts of the world for over two thousand years. It has a pleasant, mildly tart flavour, works wonders as a detox and is a great way to boost your gut microflora. Introduce it gradually: being a fairly potent source of probiotics, it's best to start out slow with kombucha. Try 30–60ml just before or after a meal at first and increase the amount when desired. As you start to feel the benefits you may end up drinking a cup or so at a time.

MAKES 1 LITRE

1 litre purified water (boiled and cooled water is fine)

1 piece of active SCOBY (you will know it's active because it will be made up of very thin layers)

2 organic or unsprayed black tea bags (don't use any flavoured or scented teas)

60g organic raw sugar (most pure sugars seem to work but don't use honey in the early days of your experimentation, as it can be less consistent)

a sterilised litre jar

food grade plastic bottle (#2 is best, #5 is acceptable; don't use any other plastic number or glass)

You will need a SCOBY (see right) or 'mother'. You can buy fresh SCOBY online or beg some from a friend who is already making kombucha. Once an obvious layer of SCOBY has developed, you can start your own batch.

Start by cleaning down your equipment and worktop reasonably thoroughly. Bring the water to the boil in a saucepan, add the sugar and stir until dissolved. Once boiling again, turn off the heat, add your teabags and submerge them. Steep for 10–15 minutes before removing the bags.

Pour the tea into the jar and allow it to cool to room temperature. This can take hours but is very important, so don't skip it.

Once the tea has cooled to room temperature, transfer the SCOBY to its new home – the top of the jar. Cover the jar with a piece of clean muslin or cheesecloth and leave in a dark corner of the kitchen at room temperature to ferment.

After about 5–7 days (depending on the taste you prefer) the first ferment of your kombucha should be about right to use. This may take slightly longer the first time you use your new SCOBY until it has grown up big and strong.

A second ferment can be carried out to add extra flavour and carbonation to your kombucha. To do this: remove the SCOBY and start a new batch. Transfer the active solution into a food-grade plastic bottle.

Add freshly juiced or cubed fruit (try using 1 apple, which will really help the fermentation) to the bottle, seal tightly and leave at room temperature for 1–2 days until you can see from its shape that the bottle is clearly under pressure from carbonation. Refrigerate and drink when desired. Use within a week to 10 days for best taste and bubbles (refrigeration dramatically slows the rate of fermentation).

Remember: fermentation is an active and powerful process, so be careful. Lids do fly off and things can sometimes (rarely) explode. Just be aware, not scared.

The secondary ferment is the perfect time to play around with flavours including herbs, like mint and lemon verbena, berries and stone fruits – whatever you love most or have in your fridge or garden.

HEALTH BENEFITS

∼ Kombucha is naturally fermented by the action of a SCOBY (Symbiotic Colony of Bacteria and Yeasts) making it probiotic-rich, which is great for your gut. These beneficial bacteria can prevent candida (yeast overgrowth) and help with improved digestion, mental clarity and mood stability.

∼ It is extraordinarily high in antioxidants, which can aid in boosting your immune system and energy levels.

∼ The drink contains glucosamines, a strong preventative and treatment for various forms of arthritis.

∼ Kombucha also contains large quantities of B vitamins, which are needed for healthy skin, nails and hair.

Shrub

Shrub is the name given to a refreshing non-alcoholic drink made by preserving fresh fruit in vinegar, with a little sugar added to offset the tartness. The best part is that you can literally use whatever fruit you have access to, so you can make shrubs whatever the season. The only rule is to ensure the vinegar is above 5 per cent acidity.

MAKES 1 LITRE

750g fruit, washed, hulled and any cores
  removed, chopped
1 litre vinegar
750g granulated sugar

2 1-litre jars

Begin by sterilising a 1-litre jar. Add the prepared fruit to the jar. Heat the vinegar in a saucepan until just before it boils. Pour over the fruit, leaving at least 2.5cm headspace. Wipe the jar and rim clean, then seal and store in a cool, dark place for 2–4 weeks, shaking occasionally.

After that time, remove the lid and strain the fruit vinegar into a saucepan, discarding the fruit. Add the sugar and bring the fruit vinegar to the boil, stirring to dissolve the sugar quickly before it boils. Remove from the heat and let it cool, then pour into another sterilised jar with a lid. Store in the fridge for up to six months. If any mould develops or fermentation occurs, discard. Use 1–2 tablespoons of a shrub in a glass of still or sparkling water or your favourite fruity cocktail.

Berry Cordial

Fresh, local berries come and go so quickly and it's always so sad when that last punnet of berries gets purchased or picked. I love to make berry cordials at the end of summer to give me a little pick-me-up in dark, gloomy weather. Perfect mixed in sparkling water or a fruity cocktail.

MAKES 1 LITRE

750g raspberries, blackberries, blueberries
  or strawberries (or a mixture)
750g granulated, raw or coconut sugar
6–8 tablespoons raw apple cider vinegar

1.5-litre sterilised bottle

Put all the ingredients into a saucepan over a low heat. Mash the berries and continue to cook gently for 10 minutes, until syrupy. Push through a sieve into another clean saucepan. Put the fruit into a small bowl with about 600ml of water. Strain that liquid into the pan with the syrup, trying to keep any seeds out of the liquid. Return the liquid to the heat and boil for 1–2 minutes, then pour into a sterilised bottle and seal. Label and store for up to three months. Once open, keep refrigerated.

Pink grapefruit Bitters

I am a massive fan of Campari. The bitter flavour has me hooked
every time. Feel free to switch the grapefruit for oranges if you
prefer a sweeter option. Hibiscus flowers and hawthorn berries can
be found online if you can't source them locally.

MAKES 500ML

1 pink grapefruit, not peeled

40g dried hawthorn
   or juniper berries

15g dried whole hibiscus flowers

4 tablespoons coriander seeds

8 star anise

2 tablespoons fennel seeds

2 teaspoons white pepper

1 teaspoon honey

250ml brandy

sterilised 500ml jar

Cut the grapefruit into squares and add to a sterilised jar, along with all
the other ingredients. Seal and shake well. Label and store in a cool, dark
place for up to six weeks to infuse. Strain into a clean jar and use 1–2
tablespoons in your drinks as a bitter flavouring.

*A really good shrub will be both tart and
sweet, which will whet your appetite while
quenching your thirst.*

ginger Beer

Ginger beer is not only delicious, it is also a great alternative to other fizzy drinks that are usually super-sugary. It reminds me of my childhood, because only on extremely hot days and special occasions were we allowed any kind of soft drinks and mum knew ginger beer was the best of its kind. She was actually more right than she knew. Ginger has a natural calming anti-inflammatory effect on our digestive system. In this recipe you first make a ginger bug, which means the ginger has been wild fermented (similar to sourdough), so it's also full of live, beneficial bacteria. Apart from being good for digestion, ginger also wards off colds and flu and still tastes like a refreshing soft drink. A great summer beverage, or indeed a mixer for your cocktails.

MAKES 2 LITRES

STAGE 1 (for the ginger bug)
2 tablespoons grated
   organic ginger
2 tablespoons organic raw sugar
250ml filtered water

sterilised 500ml jar

STAGE 2 (for the ginger beer)
2 litres filtered water
5–10cm piece of fresh ginger
175g organic raw sugar
juice of 1 lemon

sterilised 2 litre jar

STAGE 1
Put all the ingredients into a clean medium-sized jar. Place a clean muslin cloth over the top, secure with a rubber band and leave it at room temperature until active. After 4–7 days the solution should be bubbling and active, having fermented through the action of wild yeasts and bacteria present in the surrounding air. This is your 'bug'.

STAGE 2
Put the water in a large pan and bring to the boil. Meanwhile, grate the fresh ginger (use a 5cm piece for medium strength, and 10cm for a strong ginger flavour). Add the sugar to the boiling water, bring back to the boil, then add the ginger. Continue to boil over a medium heat for 15–20 minutes.

Remove the ginger with a strainer and allow the tea to cool to room temperature before transferring into a food-grade plastic bottle. Add the lemon juice to the cooled ginger tea.

Ensure your 'bug' is active, strain out all of the liquid and pour into the ginger tea. Seal the bottle tightly and leave at room temperature until the bottle bulges – this typically takes 2–3 days at room temperature. It is important you keep an eye on how carbonated your ginger beer is getting and, as soon as the bottle bulges, transfer it to the fridge. Drink within two weeks.

Citroncello

I always find that I have a glut of citrus, and not always lemons, so I thought I would try using them in my limoncello recipe, and it worked beautifully. You can switch out the below citrus with whatever you have at home. If you don't live in a part of the world where an abundance of citrus fruits grow in your garden, pick some up at the market to make this truly delicious drink.

MAKES 2 LITRES

1 litre vodka
4 lemons
2 oranges
2 pink grapefruits
1kg granulated sugar
1 litre boiling water
fresh strips of citrus peel, to decorate

2 x 1-litre sterilised jars
2 x 1-litre decorative bottles

Peel the citrus into long strips about 1cm wide. Remove any pith. Divide the citrus peel between two large sterilised jars. Cover evenly with the vodka, seal the jars and leave for a week in a cool dark place, shaking them a few times each day.

After a week, put the sugar into a heatproof bowl and pour over the boiling water, stirring until it is fully dissolved. Cool, then pour into the citrus jars and reseal. Shake and leave for another week in the same place, remembering to shake a few times a day.

After this second week, strain into decorative bottles and add a few strips of fresh citrus as garnish. Citroncello is best left to mature for a few months before drinking and will keep indefinitely.

Herby Floral Lemonade

This is a 'use what you have' recipe. Perfect for picking petals and herbs from the balcony, garden or foraging on a hike. I have included a few combinations of flowers and herbs, but feel free to experiment.

MAKES 1 LITRE

120g caster sugar
1.1 litres water
small bunch of herbs, left on the stalk in a bunch
3 tablespoons culinary-safe petals (check they are edible, unsprayed and non-toxic)
juice of 5 lemons
extra herbs and petals to decorate

Put the sugar, 100ml water, herbs and petals in a small saucepan, and bring to the boil. Allow to boil for 3 minutes, then reduce the heat and simmer for a further 5 minutes. Remove the herbs and leave the syrup to cool. Strain the lemon juice into a jug, top up with 1 litre of water and add the syrup to taste. Add some petals and herbs for decoration. Serve immediately with ice.

SUGGESTED COMBINATIONS
~ sage and rose petals
~ rosemary and lavender
~ mint and elderflower
~ thyme and orange blossom

Floral Hot Chocolate

These two hot chocolates are my favourite of a pretty bunch. You can very easily use the base recipe and play around with different herbs and edible flowers. Another nice alternative is to switch the dark chocolate for white chocolate – just use half the amount and no honey as it tends to be sweet in itself.

ROSE AND CARDAMON

SERVES 2

250ml whole milk
250ml pouring cream
150g dark chocolate
  (70 per cent is best)
1–2 teaspoons cardamom
  pods, crushed
2 teaspoons rose water
1 tablespoon honey (optional)
1 tablespoon dried rose petals,
  to decorate

Put all the ingredients except the rose petals into a small pan and heat slowly until the chocolate melts completely. Don't let the milk boil. Strain out the seeds and serve with petals as decoration.

SAGE AND LAVENDER

Follow the recipe above but use 1 sprig of sage leaves and ½ teaspoon of culinary lavender in place of the cardamom and rose water. Strain out the herbs and serve with a fresh sprig of sage and lavender.

Try orange blossom water, cloves and star anise
– or a fresh chilli split down the middle for those that like
their chocolate with a bit of a kick.

The garden

I GREW UP IN THE COUNTRY, so I was lucky enough
to have a huge backyard. We spent so much time outside as kids,
playing, running under the sprinkler, chasing the dogs, having
barbecues. It wasn't until much later in my life that gardens
became about more than play. They became places to grow food
and nurture biodiversity.

I spent some time living in Gloucestershire in England, studying
for my masters in sustainable agriculture at The Royal Agricultural
College. I moved into a granny flat on five acres that had the most
wonderful veggie patch. Roger and Angie, who owned the house,
encouraged me to grow my own food and it was then I became
truly connected to the land, the seasons, the garden, and I spent the
next two years immersed. I was so sad when my time there ended. I
moved to London, to a flat with no garden. Without the wonderful
vegetable patch, I discovered a new joy in balcony gardening.

I have since returned home to Australia, where my parents live on
a farm with 90 acres and I live, yet again, in an apartment. But I
have come to appreciate that both are equally exciting places to
grow things. I have my food garden at the farm, and my balcony
garden at home in my urban jungle. Sometimes Damien will hear
a squeal from the balcony when one of my plants flowers for the
first time. No matter where you do it, gardening and growing is a
joyous experience. You could be working on a vast veggie patch or
a herb pot on a windowsill, growing your own food, seeing the bees
buzzing in your lavender pot or a bird eating from your bird feeder.
Your garden is your own and you can make it a special place to feed
your belly, your skin and your soul.

Miniature Salad garden

A salad garden is perfect for almost anyone's home. If you have a sunny patch outdoors, big or small, there's nothing more gratifying than picking your dinner. Sad salad is not nice. You know the kind that comes in a bag from the supermarket, often dripping in unidentified sliminess. So say no to sad salad and grow yourself a salad garden. You can buy a growbag designed for tomato plants or use an old metal tub from a junk shop, a plastic builder's bucket or make yourself a simple wooden box. Drill some holes in the bottom for drainage, then fill with a mixture of organic potting compost and soil, and get seedlings of your favourite lettuce and herb varieties growing. The suggestions on the right make a wonderful mixed salad with a combination of sweet, bitter and peppery leaves.

SALAD GARDEN SUGGESTIONS

- cos or romaine lettuce
- frisée
- radicchio
- rocket
- parsley
- mint
- mustard greens
- coriander

EVERYDAY DRESSING

SERVES 4–6

30ml extra virgin olive oil
30ml balsamic glaze or,
 for a less sweet dressing,
 raw apple cider vinegar
1 tablespoon honey
1 garlic clove, grated
2cm piece of fresh turmeric
 root, grated
1cm piece of fresh ginger, grated
juice of 1 small lemon
pinch of chilli (optional)
salt and freshly ground black
 pepper, to taste

Put everything into a clean screw-top jar. Seal tightly and shake. Keep refrigerated until ready to use. This will store for a few weeks in the fridge.

Miniature Cocktail garden

Herbs are not only for salads and cooking: how about making your own cocktail garden so that you can have a spontaneous 'happy hour' whenever the mood takes you? Drill some holes in the base of a large copper pot, an enamel bowl or an old wooden box, and fill with organic potting compost and soil. Mix and match a selection of herb seedlings and create your own cocktail garden, or give it as a gift with a bottle of that special someone's favourite spirit.

COCKTAIL HERB SUGGESTIONS

- sage
- mint (try chocolate mint)
- rosemary
- basil
- lemon thyme
- lemon balm
- culinary lavender

These cocktail garden herbs would also make a great addition to your floral hot chocolate (page 104) or herbal teas.

A Few Cocktails

A gin and tonic is always great, but when fresh herbs are at hand, you can take your cocktails up a notch and feel like you're in a fancy bar, even when you're just in your garden.

BASIL AND LEMON THYME SMASH

SERVES 1

15ml sugar syrup (see method)
2 sprigs of lemon thyme, plus extra to garnish
1 bunch of basil leaves, plus extra to garnish
25ml fresh lime juice
50ml gin
ice

First make the sugar syrup. Heat equal weights of sugar and water in a small saucepan over a medium-high heat until the sugar is fully dissolved, then allow to cool. If you enjoy cocktails, it's worth making a decent amount of sugar syrup as it will keep for a long time in a sterilised bottle.

Put the lemon thyme, basil and lime juice in a cocktail shaker, and gently smash with a pestle. Add the sugar syrup and gin and top up with ice. Shake vigorously for about 20 seconds. Strain with a cocktail strainer (if you don't have one, a tea strainer will work too) into a chilled glass filled with ice. Garnish with basil and thyme leaves.

SAGE AND HONEY SPRITZ

SERVES 1

60ml honey
60ml water
45ml gin
30ml lemon juice
2–3 sage leaves
soda water
ice

First make a honey syrup by heating the honey and water in a small saucepan over a low heat until the honey is fully dissolved, then allow to cool. Set aside.

Put the gin, lemon juice, sage and syrup in a cocktail shaker and top up with ice. Shake vigorously for about 20 seconds. Pour through a strainer into a chilled glass. Add some ice and soda water. Twist the sage and use to decorate the glass.

LEMON BALM AND MINT VODKA MUDDLE

SERVES 1
for the lemon balm sugar
4 tablespoons raw sugar
8–10 lemon balm leaves
2cm piece of lemon zest

for the cocktail base
3 tablespoons water
1 lemon wedge
60ml vodka
30ml lemon juice
ice

to garnish
2 sprigs of mint
2 sprigs of lemon balm

First make the lemon balm sugar by grinding the sugar, lemon balm and lemon zest in a spice grinder until fine.

Combine 3 tablespoons of the lemon balm sugar with 3 tablespoons of water in a small saucepan. Simmer over a medium heat until the sugar dissolves. Allow to cool, then refrigerate until ready to use.

Put the remaining tablespoon of lemon balm sugar on a small plate. Wipe the rim of a cocktail glass with the lemon wedge and run the rim around in the sugar; refrigerate until ready to use.

Combine the vodka, lemon juice and 1 tablespoon of syrup in a cocktail shaker filled with ice. Shake well, then strain into the sugared glass and garnish with mint and lemon balm.

CITRONCELLO SPRITZ

SERVES 1
50ml citroncello (see recipe on page 80)
150–200ml soda water or ginger beer
1 handful lemon balm leaves
ice
thin strips of citrus peel, to garnish

Put the citroncello with the soda water or ginger beer in a large jar and add the lemon balm leaves. Pop the lid on the jar and give it a shake to mix everything together, then serve over ice. Garnish with thin strips of citrus peel, making sure the pith is removed.

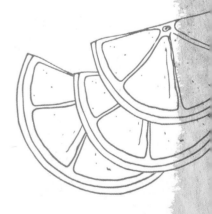

10 THINGS TO DO WITH COFFEE GRINDS

Apparently over 2.25 billion cups of coffee are drunk around the world every day. That is a lot of coffee – and a lot of waste. Obviously the good stuff is extracted in the making of your morning brew but the remaining granules are often thrown away without a second thought. It's time to rethink what would normally go in the rubbish, as there are many things coffee grinds can be used for, in the garden and beyond. Here are just a few of them.

1 - Add coffee grinds to your compost heap. Coffee
is high in nitrogen, which your compost needs.
Just don't overdo it – compost needs to be
balanced and coffee contains acid.

2 Use as bug repellents around the bottom of your
plants. The ants and snails don't like the grit,
because they hate to crawl over it. I can only
presume it scratches their little bellies.

3 It is said feral cats and strays hate coffee and that
if you put coffee grinds in areas of your garden,
the cats will not come back.

4 Put a small container in the back of your fridge
to absorb odours. Much like bicarbonate of soda,
it does a great job of absorbing smells.

5 Place in a small vase or cup in your communal
rooms as a great odour repeller.

6 Make candles with them so your house smells
like delicious coffee.

7 Use in a coffee scrub for your face and body.
The coffee boosts circulation and helps to fight
cellulite too.

8 Use them to scent your homemade soap with
them so you smell like coffee all day.

9 Use the grounds in your hair before shampooing
to get rid of any buildup – it is like an incredible
exfoliant for your hair.

10 Put a little in your vase with flowers to extend
their shelf life.

A Terrarium

Making your own terrarium is simple, and it can be done using
things from around your home. It is also very easy to maintain,
no matter how much space you have or where you live. Give it
the occasional spritz of water and you will have a terrarium for life
(almost). I often think the terrarium was created for those of us who
tend to kill all houseplants…

First find yourself a suitable container. The original terrarium was a
sealed transparent container (think miniature greenhouse) and while
glass looks pretty you can use old teapots, jars or teacups. Whether you
choose to have an open, semi-open or closed terrarium, keep in mind
that all plants have different needs. Cacti and succulents are among the
easiest because they need little attention. If you want your terrarium to
be positioned in a sunny place, an open container will work just fine,
whereas plants such as ivies that thrive in humid climates should be
placed in a closed one. Make sure you get the right soil for the plant, too,
and choose slow-growing plants to avoid overcrowding in your container.

*Do a little research on watering the particular
plant you choose, but generally speaking a
closed container almost never needs watering
and open ones need watering about twice a
week. Terrarium-grown plants will thrive
in direct sunlight, provided you give them a
spritz regularly. It is also important to prune or
remove any dead leaves often.*

Encouraging Wildlife

Encouraging biodiversity into your garden or to visit your urban balcony garden is not only important for the environment, it is also wonderful for you as well as your warm fuzzies. These two simple projects can be made using things in the kitchen or shed, as well as foraged from the park or a friend's yard. I first heard of a bug motel when I was running workshops with my dear friend and the person who I want to be when I grow up, Costa, who is a host on *Gardening Australia*. He showed a bunch of school kids how to make them and they were so excited. Now every time I do a workshop with kids, we make these two simple, miraculous contraptions that will have you counting the birds and bees in no time.

TO MAKE A BIRD FEEDER

All you need is something to keep the food somewhat protected from the weather and somewhere for the little birds to perch while they feed. After all, it's not a takeaway joint. You can use old cups and saucers, some superglue and a little rope or ribbon to tie it to a tree or post. Get creative.

TO MAKE A BUG MOTEL

You just need to make something that the little guys and gals can hide in. Twigs, branches and pieces of wood with holes drilled in them are wonderful for bugs to hibernate in. Just grab a bunch, tie them together and hang from your fence and trees. For bees you can simply take a log and drill 5cm holes into the wood.

A Wreath

Not just for Christmas, a homemade wreath is a perfect 'welcome home' hanging for your front door and a great centrepiece for a table. All you need is a base, some twine and a little creativity. You can fashion your own wreath using an assortment of branches, greenery, flowers and nuts. Using herbs is also a good idea as they look and smell great fresh and once they have dried out. The best thing is that once you have the base, you can vary the decorations seasonally to keep things interesting.

a wreath base (available online or from your local florist or garden centre)

twine

your chosen materials – you can use hardy herbs (like rosemary), homegrown or foraged branches (complete with leaves), flowers and nuts

Start by deciding if you want the main feature of the wreath to be at the bottom or perhaps the side. You can, if you choose, decorate the entire round, but personally I think leaving some space makes for a more interesting wreath and I like to position my main feature at the bottom of the wreath. Always save your hero piece until the end to make sure it's prominent.

Working from the inside out, arrange your chosen leaves and branches around the wreath, securing them with the twine. Finish with your 'hero' piece to complete it.

Use as many aromatic branches and hard herbs as you can to make the wreath smell as beautiful as it looks.

Health & Beauty

Face & Body

I HAVE SPENT YEARS speaking professionally about what we put into our body, and have always paid attention to what went into mine personally – not in an over-obsessive way, but just to ensure that all the food I eat comes from an ethical source. Recently, I realised that there is no difference between what goes into my body and what goes on to it – these products should still be ethically sourced. This realisation made me feel a little bit hypocritical, so I started looking at where my skincare products came from and playing around with recipes. For a while my home looked like a science laboratory, with jars and bottles on every shelf available. As with any shop-bought skincare product, every person's needs are different and when you first start making your own beauty products you may need to use a little bit of trial and error to get it to suit you. Just think of this as the same sort of thing as tweaking a recipe when you're cooking to suit your tastes. Use the base idea and play around with aromas and mixes. Experiment and, most of all, have fun. Some of these things may replace items in your bathroom and perhaps some will not, but the point of it is to know that there are always alternatives to mass-produced and processed products, and that you can control what you put on your body as well as what you put in it.

Like all good things, the recipes in this chapter and those following are best made in small batches and not stored for too long.

SOME TIPS

~ Anything with oil in it is best kept in an amber/dark coloured bottle or stored away from direct sunlight to avoid oxidisation.

~ Where coconut oil is used, unless otherwise specified, warm it to melt into liquid form.

~ Use raw, unflavoured honey for the best health benefits, and try to use local honey as this can help with problems like asthma and hayfever.

~ Use sterilised bottles for storage to avoid spoilage and ensure long shelf life.

~ Filtered water is advised but boiled then cooled tap water is fine.

~ Where nut oils are used, it is OK to replace the specified one with another nut oil. However, swapping olive oil for coconut oil is less likely to work unless specified, and you may need to adjust quantities in order to get the best results.

~ Where sugar is used, avoid white processed sugar as it defeats the natural purpose of the recipe. Use raw or coconut sugar – most of the time they will work in the same way.

~ When using salt, you can use any unless it has been overly processed – iodised table salt is normally processed. My favourite is Murray River Gourmet Salt, a pink salt from the Murray River in Australia, but you can use any good-quality sea or river salt. Flakes are great as they help with exfoliation. Grind rock salt a little before use as it can be extremely hard.

~ Be sure to buy or pick unsprayed herbs or flowers, otherwise there may be chemicals going into your skin after all.

~ Remember, I am not a doctor. It is always wise to test a new product on a small section of your skin (e.g. on your wrist) before slathering all over your face to test for any possible reactions. If you experience any reactions or discomfort when using these products, stop using them immediately and seek medical advice.

Anti-ageing Face Serum

MAKES 1 APPLICATION

1 teaspoon vitamin C powder
2 teaspoons filtered water

The antioxidant properties of vitamin C, and its role in repairing the collagen in our skin make this vitamin the perfect ingredient for skin health and regeneration of the cells. Serums are usually an expensive addition to your beauty regime. This homemade serum works just as well but is far more cost-effective. You can easily buy vitamin C powder in health shops or online – just make sure you buy a natural one. It's best to make this serum as and when you need it.

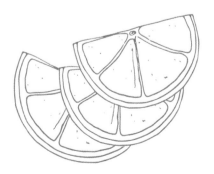

Stir the vitamin C powder into water until fully dissolved. Use before bed after washing your face and before moisturising. Smooth onto your face in a circular upward motion.

Honey and Rose Facial Toner

Honey and rose smell so beautiful together and they both have wonderful healing and anti-ageing properties. Honey works as a gentle exfoliant and is naturally antibacterial, making this toner perfect for oily skin and blocked pores. Use it after washing your face and before moisturising. I like to carry it in my handbag and use it when I need a pick-me-up or if I am travelling to stop my skin getting dry. Your skin will feel soft and smell beautiful. You can also substitute the rose essential oil for lavender or peppermint.

MAKES 200ML

1 teaspoon raw honey

10 tablespoons warm filtered water (make sure it is warm rather than boiling hot)

6–8 drops rose essential oil

1 teaspoon rosewater

1 teaspoon raw apple cider vinegar or coconut oil (optional)

lidded or spray bottle

Dissolve the honey in the warm water and mix, then stir in the essential oil and rosewater. If you want to use this as a toner, add apple cider vinegar for oily skin, coconut oil for dry skin or both for combination skin. If you are making this as a face spritz, you can leave the vinegar out.

Pour the mixture into the bottle and store in the fridge for up to three months. Shake well before each use. Apply with a cotton wool pad or directly by spray.

Eye Make-up Remover

Mirror, mirror on the wall, our eyes are the most sensitive of all. The skin around our eyes is so thin we really need to look after it. And sometimes, say after a late night out, taking off your make-up is not ideal, but this one-wipe remover will make your bedtime come a whole lot quicker. This simple recipe can also be used on your face to remove make-up pre-cleansing. Just pour a teaspoon onto a cotton pad as needed and remove make-up and grime in a circular motion, working up your face to avoid stretching the skin, and around your eyes, gently working from the outside inwards.

MAKES 300ML

10 tablespoons witch hazel (alcohol-free)

10 tablespoons your favourite oil (olive, nut, apricot kernel are all good)

sterilised glass bottle

Mix the ingredients together in a glass bottle. Shake before use. Store in a dry, dark place for up to three months.

This honey and rose facial toner is good for oily skin. If you have dry skin, add 1 teaspoon of coconut oil.

green Tea and Spirulina Face Mask

Both green tea and spirulina are thought to slow the ageing process, so pile this on your face as often as possible – and enjoy your daily cup of green tea while you're at it. You can use peppermint or chamomile tea if you don't have green tea.

MAKES 1 FACE MASK
1 green tea bag
2 teaspoons spirulina powder

Make yourself a cup of green tea and allow to steep for 4–5 minutes. You only need 6–8 drops of the tea, cooled, so drink all of it except that amount. Mix the spirulina powder and tea to make a paste. Smooth onto your face and leave for 20 minutes. Wash off with warm water and a flannel.

Sugar Face Scrub

There is no better feeling than a truly good scrub after a long day. Exfoliants can be overly harsh on your skin – especially your face. A lot of shop-bought scrubs contain microbeads, which, frankly, are dreadful for the environment. This scrub is gentle and does no harm to the environment at all – plus you could probably make it from what you have in your cupboard now. You can triple this recipe and use all over your body too.

MAKES ENOUGH FOR A SINGLE USE
4 teaspoons organic granulated coconut sugar
4–6 drops filtered water
2–4 drops orange, lemon or lavender essential oil

Combine all ingredients in a small bowl. Using the tips of your fingers, gently rub scrub into your cleaned face, taking care to avoid your eyes. Remove by rinsing with cool water, and pat dry with a clean towel. Use every few days so that you don't over-exfoliate.

Add a tablespoon of coffee grinds to your scrub and an extra drop or two of water, then triple the quantity to make a coffee body scrub that's great for cellulite and a fantastic exfoliator.

Blueberry and Honey Mask

Blueberries are filled with anti-oxidants, honey is antibacterial and you don't have to eat them to get the benefits. This super-simple mask means they go directly into your skin. Try sourcing local honey for added health benefits, too, such as relieving hay fever. Ideally, you want fresh blueberries, but you can also make this using frozen and thawed fruit.

MAKES 1–2 APPLICATIONS

100g fresh blueberries
2 tablespoons raw honey
2 tablespoons raw or
   coconut sugar

Put all the ingredients in a small food processor and blend until smooth. Add a generous layer to your clean face, and let it soak into your skin for 15 minutes. Rinse clean with warm water. I would recommend making this and using it straight away or within a few days. Store in the fridge if not applying immediately and use once or twice a week. You may want to use this as an excuse to lie down for 15 minutes as it can drip down your face.

The best part is that this mask is both edible and delicious. If you take a lie down while you're waiting for the mask to work its magic, just open your mouth and anything that happens to fall in is an added bonus.

Macadamia and Vanilla Body Butter

If you don't want to eat this after you have made it I will be surprised! Thick and luscious, you will delight in the silky-smooth feel of this butter on your skin.

MAKES 250ML

60ml shea butter

60ml cocoa butter

60ml macadamia oil (or any light nut oil such as almond or olive oil)

4 tablespoons coconut oil (warmed if it has set)

10 drops vanilla essential oil or the seeds from 2 vanilla pods

Combine all the ingredients except the essential oil in a double boiler and gently heat. Alternatively, place in a heatproof bowl over a saucepan of just simmering water. Stir until the ingredients are fully melted, then stir in the essential oil.

Remove and allow to cool until cool enough to touch. Place in the fridge for 30 minutes to an hour or until the butter is starting to harden on top, but still soft.

Mix with a hand mixer or stand mixer for about 8–10 minutes or until light and fluffy.

Store in a jar or airtight container and keep in a place where the temperature is no higher than about 25°C; any hotter and it may melt or, if too cold, it will solidify. (If you live in a hot climate, store in the fridge.) Use within two months and use as often as you wish.

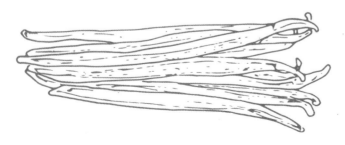

*The vanilla is added just
for the scent. You can switch
it out with whatever essential
oil you love most.*

Bath Salts

Nothing beats a good soak in a warm bath and Epsom salts (magnesium sulphate) have been used for their therapeutic effect on aching joints and muscles for over a century. Used with potassium- and iodine-rich sea salt, these minerals are absorbed directly into the skin, helping with dermatitis and inflammation. Adding essential salts relieves stress and works to improve your overall well-being.

LAVENDER AND PEPPERMINT BATH SALT SOAK

Both these essential oils promote relaxation so it is best to enjoy this soak before bed or at the weekend when you are lazing about.

MAKES 700G
(enough for a week of baths)

600g Epsom salts
75g sea salt
160g bicarbonate of soda
20 drops lavender essential oil
10 drops peppermint essential oil

Mix all the ingredients into a small bowl and store in a jar or airtight container for up to three months. Pour about 100g into the bath while the water is running. Use as often as needed.

ROSEMARY AND LEMON BATH SALT SOAK

Rosemary is said to restore strength and reduce anxiety, while the lemon is reviving so together they make a great anti-stress and rebooting soak.

Follow the recipe on the left but replace the essential oils with 20 drops of lemon essential oil (or the zest of a lemon) and 1 finely chopped sprig of rosemary.

For a truly luxurious bathing experience, sprinkle flower heads into the bath.

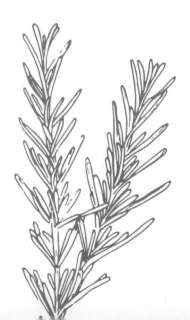

Bath Fizzes

Bath fizzes, sometimes referred to as bath bombs, are a wonderful homemade gift. They are simple to make and last for over a year in an airtight container. They also look pretty displayed on the side of your bath and make bathing fun, as well as giving you the benefit of the salts and oils.

MAKES 2–4 FIZZES
DEPENDING ON SIZE

50g arrowroot flour

115g bicarbonate of soda

50g citric acid

50g Epsom salts

1–1½ teaspoons water

1 teaspoon essential oil
  (choose your favourite)

1¼ teaspoons olive oil

1–2 drops natural food colouring
  (see page 92)

sphere moulds or ice-cube trays
  (silicone is best; you can buy
  bath moulds online)

Combine the first four ingredients in a bowl. Stir and remove any clumps with a fork or hand whisk. Put the rest of the ingredients in a small glass jar and stir together. Slowly pour the liquid mixture into the dry mixture, whisking as you go. Be patient – the mix will foam if poured too quickly. Once the wet ingredients are combined with the dry ones, take a small amount in your hand and squeeze it together. It should stick together well in one to two big pieces. Add a little more water if it does not stick and try again.

Fill each half of your sphere moulds with the mixture until it's just above the rim of each mould. Press together. If using an open mould, just fill and leave. Allow the mixture to dry out for about 10 minutes before removing one half of the mould. Gently invert the bath fizz so the uncovered side is facing down, then carefully remove the other half of the mould from the top. These spheres can fall apart very easily if you're not gentle.

If a bath bomb does crack in half while you're removing it, you can gently press it back on top of the other half, or return the mixture to the mixing bowl and start the moulding process again. Once they are completely out of the moulds, allow them to dry out overnight.

Store in a dry place or airtight container until you're ready to use them. Drop into the bathtub while the water is running and use 1 or 2 at a time.

Use a mixture of pastel colours and fill a pretty jar,
adding some dried petals or lavender, to make the perfect gift.

Massage Oil

Perfect for easing an aching body or making you feel relaxed and uplifted. Who wouldn't want to be massaged? Make it for your love as a present and ask them to test it on you…

MAKES 250ML

120ml grapeseed, olive or
  light carrier oil
120ml coconut oil
  (warmed if it has set)
35 drops eucalyptus essential oil
35 drops lavender essential oil
22 drops peppermint essential oil
20 drops ginger essential oil

amber- or dark-coloured bottle

Pour the two oils into a clean bottle using a small funnel. Add the essential oils. Screw on the lid and shake. Store in a cool, dark place for up to a year.

This also makes a fantastic bath oil.

Body Scrubs

ROSEMARY AND CITRUS SALT SCRUB

Salt (in moderation) is fantastic for your body, both inside and out. Avoid iodised salt, as it has been processed. Good-quality salt is full of the minerals magnesium, potassium, sodium and calcium, which are fabulous for cell regeneration. Using salt directly on your skin helps to protect it from inflammation and breakouts, as well as balancing oil production. This is my favourite salt scrub – rosemary and citrus make an invigorating combination.

MAKES 450G

220g good-quality salt
230ml light oil (not olive), such as jojoba or nut
10 drops orange or lemon essential oil
  or half and half
1 sprig of rosemary, finely chopped

Mix the salt and oil in a small bowl until it holds together. You don't want it to be too runny, so add more salt if it feels this way. Add the essential oils slowly and then mix in the rosemary. Store in a jar or airtight container for up to three months. Scoop up a handful of the scrub and use on damp skin, using light, circular strokes, then shower off. Avoid using on broken or irritated skin and don't use this on your face or near your private parts.

BANANA AND CINNAMON SCRUB

Use up those old bananas in more than just a cake. You will, of course, smell like a banana cake but hey, who doesn't love cake? Play around with essential oils to your liking, but I think banana and cinnamon are the best of friends.

MAKES ENOUGH FOR A SINGLE USE

1 super-ripe banana
4 tablespoons raw or coconut sugar
4–6 drops cinnamon essential oil or 1 teaspoon
  cinnamon powder (oil is a better option)

Loosely mash the banana in a small bowl. Mix in the sugar and oil. Use before you shower by smoothing onto your skin (not face) then give yourself a well-deserved massage, savouring the smell. Turn on the water and rinse.

COCONUT, HONEY AND MINT FACE AND BODY SCRUB

The coconut helps with moisturisation, honey has antibacterial and anti-ageing properties and the mint is anti-inflammatory. Used a couple of times a week, this scrub will give your body a little extra love.

MAKES 125G

125g organic coconut sugar
1 tablespoon coconut oil (warmed if it has set)
1 tablespoon raw honey
¼ teaspoon lemon juice
½ handful spearmint or peppermint leaves, ripped
  into tiny pieces

Combine the sugar, oil and honey in a small bowl, then mix in the lemon juice and mint. Store in a jar or airtight container for two weeks. Make it in small batches so that it is fresh and active when you use it. Scoop up about a tablespoon at a time and rub onto your skin in the shower or bath, rubbing in a circular motion. You can use this as often as needed.

Foot Scrub

Salt is perfect for relaxing the muscles and treating cramp. Combined with the rejuvenating properties of peppermint and lavender this scrub is the greatest treat for tired feet. Use after a hard day at work or as often as needed. Magnesium flakes are incredible for aches, pains, inflammation and muscle spasms.

MAKES 300G

300g Epsom salts or magnesium flakes
50ml olive or nut oil
1 teaspoon liquid castile soap
10–15 drops peppermint or lavender essential oil
  (or mix of the two)

Mix the salts with the oil and soap in a small bowl, then add your chosen essential oil. Store in a lidded jar or airtight container, and use a teaspoon or so at a time to exfoliate your feet – or body – as needed. Rinse well after use.

Deodorant

This may or may not work for you. Sadly, I suffer from smelly underarms so tend to buy natural deodorant instead of making my own but some of my friends have had much success with this and similar recipes. Lucky you who smell like roses… Give it a try. Every person is different.

MAKES 150G

50g bicarbonate of soda
50g arrowroot powder
10 drops essential oil (try tea tree, lavender
  or rosemary)
4 tablespoons coconut oil (warmed if set)

Mix the bicarbonate of soda, the arrowroot and the essential oil to make a smooth paste. Begin to incorporate the coconut oil, a little at a time, until the desired consistency is reached (I would suggest a paste thick enough to spread on your skin). Store in an airtight container to stop it from drying out. Use by wiping a little with your fingers on your clean armpit.

Peppermint essential oil is helpful in purifying the skin and increasing circulation, so is great for tired feet. Its cooling effect also helps you feel refreshed. Tea tree has natural deodorising and antifungal properties so is great to add for stopping that dreadful smelly feet feeling.

Oral Hygiene

TOOTHPASTE

Commercially made (non-natural) toothpaste contains artificial gels, dyes and a very long list of chemicals. I am not going to lie; yes, it leaves my mouth feeling super clean and it's hard to match, but give this recipe a go and see how you feel. I am no dentist and I do not know the long-term benefits or effects, so I suggest you do a bit of research yourself and see what your dentist says about it next time you're in the chair.

MAKES 120ML

8 tablespoons coconut oil (solid)
2 tablespoons bicarbonate of soda
20 drops peppermint essential oil

Put the coconut oil in a small bowl and work it with a spoon to soften. Gradually add the bicarbonate of soda and mix it thoroughly. Add the essential oil and mix again. Put the mixture into a sterilised jar. Use the same amount as normal toothpaste by putting a small amount on your toothbrush using a clean spoon. Brush and rinse as normal. Keep for up to a month.

MOUTHWASH

As with commercial toothpaste, I am super-paranoid about swallowing the chemicals contained in most mouthwash brands. Don't swallow this homemade one either, but at least if you do accidentally ingest some of it, it's not nearly as bad.

MAKES 250ML

250ml filtered or distilled water
2 tablespoons bicarbonate of soda
2 drops tea tree essential oil
2 drops peppermint essential oil
1 drop cinnamon essential oil

Mix all the ingredients in a small jar. Shake well before each use and keep for up to two weeks. Best made regularly. Avoid swallowing.

Adding a few drops of clove oil will soothe mouth ulcers or tooth aches.

SHOP-BOUGHT HAIR PRODUCTS can be really expensive and more often than not they are full of silicones. While silicone originates from a natural source, it undergoes extensive chemical processing that leaves a big ecological footprint and makes the end ingredient synthetic. Unless you have time to read every single label top to bottom and know what every ingredient is, buying ethically from a shop can be really hard work and incredibly frustrating. These homemade alternatives will not only save you money, they will encourage strong, shiny hair and give you a healthy, chemical- and plastic-free shower. Don't be put off by the thought of putting ingredients like raw egg or oil on your hair because they will wash out when rinsed properly – you'll not be left with scrambled egg in your hair or greasy locks if that's what you're thinking. These are just great ways to use natural, inexpensive ingredients you are bound to have at home. Follow the instructions and give it a go, even if it's only every now and then. It's worth bearing in mind that it took me quite a few times using these recipes to notice a difference because I had a build-up of all kinds of stuff in my hair, so you need to be a little patient when trying these out.

Hair &
Make-up

Coconut Hair Care

COCONUT AND ORANGE SHAMPOO (FOR NORMAL HAIR)

Coconut is the perfect moisturiser and will leave your hair silky smooth, while the orange essential oil makes you smell good enough to eat. Who doesn't want that?

MAKES ABOUT 120ML
(enough for about four washes, depending on the length of your hair)

60ml coconut milk, canned or handmade*
60ml liquid castile soap
20 drops orange essential oil
½ teaspoon olive or nut oil (optional, for dry hair)
*as canned coconut milk tends to separate in the can, you can either just use the liquid part or, if you prefer a thicker consistency, shake the can well to mix the solid and liquids and then measure out your 60ml

Combine all ingredients in a recycled shampoo bottle or an airtight container or screw-top jar. Shake well to mix. This is fine to store in your shower for up to a month, just remember to shake well before every use.

COCONUT CONDITIONER

This gorgeous conditioner will leave your hair doubly smooth. First use pure coconut oil to moisturise and condition, then follow with the 50:50 mixture of apple cider vinegar and water as a rinse to really clean your hair and help with scalp problems such as dandruff. The acid removes any excess oil from your hair and increases its shine. Try using it once a week or so.

MAKES ONE APPLICATION
1–2 teaspoons coconut oil
60ml raw apple cider vinegar
60ml water
10 drops essential oil: sage (for normal hair), tea tree (for oily hair), or lavender (for dandruff)

Make sure the coconut oil is liquid – place in a warm place to melt if need be. Put the vinegar, water and essential oil in an airtight container or screw-top jar. Shake well to mix. You can make a larger quantity of the vinegar rinse and keep it in your shower for up to a month – just remember to shake well before every use.

Start by rubbing the coconut oil into the ends of your hair and leave for a minute. Rinse thoroughly in warm water. Pour the vinegar mix through your hair and rinse thoroughly.

*Everyone's hair is different, so you may need
to adjust these recipes and play around with them a few
times before you get the perfect balance for your hair.*

Hair Masks

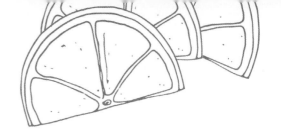

All of these storecupboard staples make for perfect hair masks. Wash your hair and towel dry. Use any of the following by applying directly to your hair and rubbing it in. Wrap your hair in a towel or put on a shower cap and leave for at least 10 minutes or up to an hour. Rinse well. Measurements below may vary depending on hair thickness and length, so you may need to use more or less than suggested.

EXTRA VIRGIN OLIVE OIL

Olive oil has been used for health and beauty for centuries, and is still used for these purposes in many Mediterranean countries today. It's a good choice for a conditioner as it hydrates your hair – great for dry and brittle split ends.

Between 60–120ml

HONEY

Our bees deserve a high five. Honey contains vitamins, minerals, it's antibacterial, acts as a humectant (retains moisture), and has wonderful healing properties. It's great for everything, including your hair. Used as a mask, it adds moisture and shine.

Between 2–6 tablespoons

LEMON

I used to use lemon as a bleach as a teenager when mum wouldn't let me dye my hair. Two decades later, here I am doing it all over again. Lemon is as good as bleach and can naturally lighten hair if used regularly enough. If you do not want to lighten your hair but want to get rid of any excess oil as it closes pores in your scalp, just dilute the lemon juice 50:50 with water.

2–4 lemons

EGGS

Jam-packed with protein, eggs are great for encouraging hair growth. Use as a rinse or mask from time to time. Remember not to use super-hot water or your hair will end up with scrambled eggs in it. Beat the egg(s) and apply to wet hair. Leave for 20 minutes and rinse in cold water. Wash as normal.

Between 1–3 eggs

COCONUT OIL

Full of fatty acids, coconut oil makes your hair soft and silky smooth and promotes natural shine. Just avoid it on your scalp, as this is already the most oily part of your hair, and remember to rinse thoroughly. Best left in overnight as a treatment.

Between 1–4 tablespoons

Dairy Hair Treatment

Treatments are really expensive. This simple recipe, using things you already have in the fridge, is beyond cheap and will leave your hair feeling like you just stepped out of the salon. Your hair is basically made up of proteins, as are dairy products, so this will encourage thick, strong, healthy hair.

MAKES ENOUGH FOR ONE APPLICATION

1–2 eggs (1 for short hair, 2 for long hair)
125–250ml whole milk (125ml for short hair, 250ml for long hair)
2–4 teaspoons olive oil (2 for short hair, 4 for long hair)
1 teaspoon lemon juice

If you have normal hair, use the egg(s) whole. If your hair tends to be oily, use egg white only or, if your hair is dry, use egg yolk only.

Choose your egg accordingly and beat with a whisk. Add the milk and oil and whisk again, then mix in the lemon juice.

Massage into your scalp and work it through to the ends of your hair. Put on a shower cap and leave for 10 minutes. Rinse with cold water (lukewarm if you're not so brave). Rinse thoroughly and shampoo or use an apple cider rinse. Use once a week for best results.

Detangler

I still remember the feeling as a young girl of wriggling and whingeing as my hair was brushed after being washed. It was and still is to this day prone to knots and tangles. I wish mum had had this detangling treatment to hand back then – it may have made for easier bath times.

MAKES 200ML

200ml distilled water
2 tablespoons raw apple cider vinegar
6–8 drops lavender essential oil

200ml spray bottle

Using a small funnel, pour the vinegar and essential oil into a recycled spray bottle. Add the water, screw on the lid, shake and use as necessary by spraying into the ends of your hair and combing through. Store in your bathroom for up to a year.

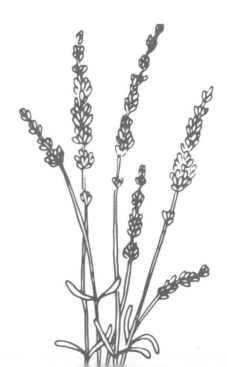

151

Essential Oils for Hair

Essential oils are wonderful – and not just for their vibrant smell. They each have healing properties and are great for targeting specific ailments. It is really important to do your research as to what is suitable if you are pregnant or are suffering from any illnesses, allergies or disease, because essential oils are not suitable for certain conditions. Please make sure you check carefully about any that are to be avoided prior to using.

These oils are listed for you to use in place of others in the various hair recipes, depending on your specific needs.

FOR HAIR GROWTH
Rosemary

FOR DANDRUFF
Lavender

FOR STRENGTHENING
Jojoba

FOR DRY AND DAMAGED HAIR
Basil

FOR ITCHY AND SORE SCALPS
Tea tree

Salt Spray

Perfect for that 'just been for my morning swim' look, without the nasty chemicals. This will work wonders for adding texture and curl to your hair. Spray and scrunch on towel-dried hair for best results. This is a great way to achieve a styled look without the chemicals.

MAKES 250ML
250ml hot (not boiling) water
2 teaspoons sea salt
1 tablespoon coconut oil, macadamia oil, almond oil or argan oil

250ml spray bottle

Mix the ingredients together and pour into a spray bottle. Shake well before each use. This will keep for up to three months.

Natural make-up

Walking into a shop and looking at shelves and shelves of make-up products can be really quite overwhelming, especially when you are trying to make ethical and natural choices. Products filled with E-numbers, plastics, silicones and who knows what else fill every available shelf, and for the most part we shop mindlessly, dumping item after item into our baskets, paying through the roof for chemical-laden and artificial beauty products. Making your own make-up is as good as feeding yourself healthy, fresh and unsprayed food. It all goes into your body after all. It can also be so much cheaper in the long run.

My interest in make-up started pretty young. My earliest memories are of me rummaging through my mum's make-up bag and plying my face with blue eyeshadow and red lipstick. Classy indeed, but hey it was the 80s after all. My auntie Sarah often tells the story of my obsession with flushing her make-up down the toilet. Maybe even back then I knew it was full of chemicals.

I used to wear a lot of make-up as a teenager and remember having awful skin because my pores were constantly clogged with thick layers of foundation and powders. None of the organic and mineral types were available then, and it was not something we knew much about. In fact, it's really only been in the past five to ten years that the natural make-up industry has become such a key part of the marketplace. Even now, making the switch can still be confusing and expensive. The recipes I have included here are really just a few simple and cost-efficient ones for you to see how you feel about them before investing in more expensive items or ingredients. Most importantly, as I keep saying, if it shouldn't go in your body, it isn't in this book.

Concealer

MAKES 40ML

1 teaspoon shea butter
2 teaspoons jojoba oil
½ teaspoon emulsifying wax (used in beauty products to help it mix together and stick)
1 tablespoon aloe vera gel
1 teaspoon witch hazel
½ teaspoon cacao powder
½ teaspoon natural clay (sold as powder – choose colour based on skin tone)

Melt the shea butter, jojoba oil and emulsifying wax in a double boiler or even a jam jar in a pan of simmering water until completely melted. Add the aloe vera gel and witch hazel and whisk until completely incorporated and smooth. Remove from the heat.

Add the cacao powder and clay a little at a time until you reach the colour you desire. Once cool, test the colour and coverage on your cheek to make sure you've achieved the right tone for your skin and the amount of coverage you want. Spoon the mixture into a small airtight container and use as needed. Store for up to three months.

All of these recipes are best made fresh in small batches and not kept for too long.

Face Moisturiser

This is a really great daily moisturiser to use under an SPF-based make-up or at night. It also makes a great base for my tinted moisturiser recipe (see right), but you will need to supplement it with an adequate SPF product. When it comes to daytime wear, I am very particular about using sun protection as a few years ago I had a skin cancer removed from the corner of my eye. So I normally buy an SPF face moisturiser for the daytime to help my skin cope with the harsh Australian climate.

MAKES ENOUGH FOR A MONTH

6 tablespoons aloe vera gel

3 tablespoons coconut oil

1 teaspoon vitamin E oil

2 tablespoons sweet almond oil

Mix the ingredients together and store in a sterilised bottle for up to six months. Use once a day.

Tinted Moisturiser

These days I actually don't often wear much makeup, but I do tend to wear a tinted moisturiser most days. This combines some natural powders with the above moisturiser recipe so you can make it to suit your skin depending on the season.

It will last for a month or two, depending on how liberally you use it. Quarter the recipe to make smaller batches if you like, and keep it in the fridge for the best shelf life.

MAKES ENOUGH FOR A MONTH

6 tablespoons natural face moisturiser (see right)

½–1 teaspoon mica powder in beige, bronze or pink, depending on what suits your skin

1½ teaspoons cacao powder

Mix the ingredients together little by little until you get the desired shade to suit your skin. Store in an airtight container for up to a month.

No matter where you live, sun protection is important.
The harsh reality of climate change means that we all really should
be conscious of our environment. I don't ever go outside without sun
cream on, especially on my face. I would suggest investing
in a natural SPF product to add to your routine.

Blusher

My brothers have always teased me about how much blusher I wear. I love it. There is no particular reason, it is just the one thing I always wear and that will never change. It takes me from pasty white girl to pasty white girl with happy cheeks.

MAKES ENOUGH FOR
A MONTH (unless you use it
as much as me)
½–1 teaspoon cacao powder
   (depending on skin tone)
1 teaspoon mica pigment or
   beetroot/hibiscus powder
1 teaspoon arrowroot powder

Mix the ingredients together, blending in the cacao a little at a time so that you get a good colour for your skin, and store in an airtight container. Apply with a blush brush – you may need to play around with quantities depending on skin colour.

This is my favourite of the make-up recipes because it makes my cheeks smell like chocolate. Petal powders will be grainy in comparison to mica powder and will look a little glittery. Use the Mica powder for a smooth blush.

Mascara

Mascara is one of those things that can, in an instant, transform your entire face. But anything that requires chemicals to scrub it off at the end of the day surely can't be great for your eyes, can it? This natural version is a safer bet.

MAKES ENOUGH FOR 2 WEEKS

4 teaspoons coconut oil

8 teaspoons aloe vera gel

2 teaspoons grated beeswax

2 teaspoons activated charcoal (for black) or cacao powder (for brown) or mica powder for coloured mascara

an empty mascara container

Put the oil, aloe vera gel and beeswax into a small saucepan over a low heat. Stir until the beeswax is completely melted.

Tip the charcoal, cacao or mica powder into the mixture. Stir until combined and remove from the heat. Use a piping bag or teat pipet to push the mixture into the mascara tube. Place the wand in and seal. Store for up to three months.

Rose Lip Balm

This is a really lovely recipe and makes a delightful little gift. So easy to make and you can play around with smells and colours just using the base recipe. Try using vanilla or orange essential oils instead of rose, and take out the powder for a clear balm.

MAKES 25ML

4 tablespoons coconut oil

2 tablespoons shea butter

2 tablespoons beeswax

1 teaspoon rose powder (see tip, or you can use coloured mica pigment)

3–4 drops rose essential oil

Melt the oil, butter and beeswax in a double boiler. Stir until fully melted. Remove from the heat and stir in your rose powder or coloured mica. Now stir in your essential oil. Use a teaspoon to transfer into a small tin, smooth and let cool.

Make your own rose powder by drying out edible unsprayed rose petals and blitzing them into a fine powder in a food processor.

For Men

MY PARTNER, DAMIEN, is a well-groomed, dashingly handsome man. Our bathroom is full of all of the creams and soaps, balms and oils imaginable. It is fair to say he spends more time getting ready than I do. He will not like me saying so, but it's true. I admit, at times it infuriates me – but then I wonder, would I rather have a man who doesn't care what he looks like before leaving the house? No, I think not. Thank you for being my guinea pig for these products, Damien. Your skin is thanking you – as is your wallet.

Hair Come Back (Hair Spray)

A receding hairline can be a problem for many men but it's not without remedy. This 'hair come back' spray is natural and has been said by my testers to make their hair 'appear' thicker after time. You may not end up like Rod Stewart, but what have you got to lose? Hopefully not more hair!

MAKES 500ML

500ml filtered water

6 tablespoons dried nettle leaves

20 drops white sage essential oil

20 drops rosemary essential oil

20 drops lavender essential oil

sterilised 500ml spray bottle

Boil the water in a pan, turn off the heat and add the dried nettle. Let the herbs steep for at least 20 minutes or until the water cools. Strain the herbs through a sieve and pour the herb-infused liquid into a sterilised spray bottle. Add the essential oils and shake well. Store in the fridge for up to three months and shake well before use.

Facial Hair

BEARD OIL

Not just for hipsters! This oil is great for keeping that man beard of yours in check. The combination of oils is great for preventing dandruff, reducing itchiness and promoting growth, as well as giving the hair a great shine.

MAKES 110ML

45ml almond oil

45ml jojoba oil

1 tablespoon coconut oil (melted)

5 drops cedarwood essential oil

5 drops sandalwood essential oil

Use a small dark-coloured glass bottle with an eyedropper if you have one. Pour in the jojoba, then the almond and coconut oils. Shake, then add the essential oils. Store in your bathroom cupboard for up to a year. Use daily by putting a few drops into your palms and massaging through your beard.

COCONUT SHAVING CREAM

For those old-school lads who still shave at home – this cream is for you. It will leave your freshly shaved face feeling silky smooth and the inclusion of coconut oil will stop any ingrowing hairs or bad bacteria.

MAKES 350ML

170ml shea butter

170ml coconut oil

2 tablespoons olive oil

2 tablespoons liquid castile soap

20 drops lavender essential oil

Melt the shea butter and coconut oil in a small saucepan over a low heat, stirring until fully melted. Add the olive oil and stir until mixed. Remove from the heat. Pour the mixture into a bowl and place in the fridge until it becomes solid.

Using a handheld mixer, whip the mixture until fluffy – this takes about 2–4 minutes. Add the castile soap and blend again. Finally add the essential oils and whip until fully blended and fluffy. Scoop into a small tin and store in the bathroom cupboard for up to three months.

Chocolate and Coffee Man Scrub

Man skin can often be a little rough. If you use this a few times a week, your skin will be noticeably silkier. Cacao is rich in antioxidants and the coffee helps to exfoliate. Plus the smell is outrageously good.

MAKES 125G

125g organic coconut sugar
1 tablespoon coconut oil
1 tablespoon cacao butter
1 tablespoon honey
1 tablespoon coffee grinds

Mix all of the ingredients in a small bowl with a spoon. Store in a jar or airtight container for up to two weeks. Make it in small batches as needed as it's best used fresh and active. Apply to wet skin over the entire body, massaging into the skin and rinsing with warm water.

I sometimes make this scrub for my partner Damien, but invariably end up using it myself because it smells so good and works like a charm.

Man oh Man Body Butter

What could be nicer than silky smooth man skin that smells good enough to give an all-day hug to? I can't think of anything. Man oh man.

MAKES 220ML

60ml shea butter
60ml cocoa butter
60ml sweet almond or macadamia oil
60ml coconut oil
10 drops frankincense or sandalwood essential oil

Combine all the ingredients except the essential oil in a double boiler or a heatproof bowl over a saucepan of simmering water. Bring up to a medium heat slowly and stir until the ingredients are fully melted. Remove and leave until cool to the touch.

Pour the mixture into a bowl and place in the fridge for up to an hour or until it's starting to harden on top, but still soft. Mix with a handheld mixer for about 8–10 minutes or until light and fluffy. Finally, add the essential oils and whip until fully blended and fluffy.

Store in a jar or airtight container and keep in a place where the temperature is no higher than about 25°C; any hotter and it may melt or, if cold, it will solidify. (If you live in a hot climate, store in the fridge.) Use within two months and use as often as you wish.

Achey Breaky Balm

Men and sport tend to go hand in hand and often you don't know when enough is enough, coming home from a game bumped and bruised. This ache balm is a much less potent and chemical-free version of tiger balm, but just as effective. Rub a small amount into your fingertips, then massage into the affected area. Use as needed.

MAKES 60ML

15g beeswax
50ml coconut oil
10 drops eucalyptus
5 drops peppermint

Put the beeswax and coconut oil into a double boiler or a heatproof bowl over a saucepan of simmering water. Bring up to a medium heat slowly and stir until the ingredients are fully melted. Remove from the heat. Mix in the essential oils and pour into a small storage tin. Allow to cool fully and put on the lid. This should last for six months.

167

Remedies

I THINK OUR MODERN RELATIONSHIP with antibiotics is problematic. I realise that they save lives in some instances and that they are necessary for serious issues. However, they are not candy, and should not be handed out as such. I am extremely disappointed by report after report stating our growing immunity and almost addiction to them. I personally will try every other available option before committing to a course of antibiotics. If you do need to take them, at the very least try to take some good probiotics alongside them, like sauerkraut (page 70) or kombucha (page 96) to help build your gut bacteria. The following remedies are not going to substitute medication if you need it. I have done a course in herbal medicine, but remember I am not a doctor – you must see an expert in case of illness. These recipes will, however, help build immunity. They may also help a sore throat or make you feel a little spring in your step during flu season. They are age-old remedies with a little modern twist and are meant, like all of these recipes, for you to try at home and see how you go.

May the vinegar be with you (not the antibiotics).

SOME SUGGESTIONS

~ Use raw apple cider vinegar for best results (look for it in the refrigerated aisle, as this generally means it hasn't been pasteurised).

~ Dried herbs in tinctures and tonics are more potent than fresh herbs, so if you dislike super-strong flavours, adjust quantities to suit your taste.

~ Always use sterilised jars and bottles to avoid any cross-contamination.

~ Use raw honey for the health benefits associated and try to use local honey where possible to help with any hayfever or asthma.

~ All of the spices and herbs I have suggested are available online. What you have access to in your own country may vary, so please check carefully before buying or picking something that looks similar (it is not fun if you have a reaction or eat something poisonous).

Medicinal Vinegar

Preventative medicine to me is taking medicinal vinegar once a day. Herbs have been used for centuries to improve our health and well-being, and here they are combined with raw apple cider vinegar, which has also been said to help with weight loss, stabilise blood sugar and absorb nutrients. This combination creates medicinal vinegar, a really effective way to take your preventative medicine daily. Choose the right herb to help with your particular ailment or concern (see herb guide on pages 174-5), then follow this recipe. Remember to always check botanical names to make sure you are definitely using edible varieties.

MAKES 250ML

50g dried or 50g fresh herbs for a
 lighter version
250ml raw apple cider vinegar

250ml sterilised glass jar with
 screw-top lids

Use a pestle and mortar to crush your dried herbs to a coarse powder or chop fresh herbs. Add your herbs to a sterilised jar. Pour over your apple cider vinegar until the jar is filled to the top. Screw on the lid and store in a cool, dark place for two weeks to allow the herbs to infuse. Shake the jar daily.

Once it has sat tight for a fortnight, strain the herbs through a fine sieve or muslin/cheesecloth into another sterilised jar. Allow it to sit for a further two days so the sediment can settle and then decant the clear liquid layer into another sterilised jar using your strainer. Put the lid on and store for up to six months in a cool, dark place. I would suggest keeping in the fridge.

Herb guide

There is an incredible array of herbs available to assist in your health, wellbeing and culinary endeavours. These are just a few. The greatest pleasure I have found in my career to date is blending herbs, not just for flavour but for health. The most important thing to remember is to do your research – don't take risks when picking wild herbs, toxicity is real and it's important to know what is safe as well as what you are sensitive and allergic to. Then you can experiment and most of all have fun.

ANGELICA
Great for irritable bowel syndrome (IBS), gas and anything to do with your digestion. It also acts as an anti-inflammatory, so it can help with arthritis.

BURDOCK
A blood cleanser, diuretic and often used to treat eczema, arthritis and gout.

CALENDULA
Can be used internally and externally – wonderful for rashes, wounds and burns. Helps with ulcers, reflux and is antimicrobial.

CHAMOMILE
Well known for its ability to assist in sleep and can also help with anxiety and stress.

CHICKWEED
Acts as a blood cleanser and is good for iron deficiency.

DANDELION
A perfect all-rounder and widely available. The root encourages a healthy gut and can stimulate digestion.

ECHINACEA
Wards off colds and flu. It stimulates our immune system so we can fight infections quicker.

GARLIC
A potent tonic for boosting circulation and fighting off colds and flu. Best to avoid if you suffer from heartburn or are prone to gas.

GINGER
An amazing all-rounder for motion sickness, nausea, digestion, infection and to boost circulation.

HIBISCUS OR ROSELLE
Has antimicrobial properties and helps with high cholesterol and lowering blood pressure. Also great for the liver and kidneys.

HOLY BASIL OR TULSI

A remedy for colds and flu, sinus infections and can also help with anxiety, asthma and coughs. It is also said to be beneficial for aiding concentration and poor memory.

LEMON BALM

Helpful for depression and anxiety and also wonderful for unwinding and aiding in inducing sleep. Mild enough for children and elders.

LEMONGRASS

Helpful for headaches, stress, coughs, indigestion and menstrual cramps.

LEMON VERBENA

Good for nausea and insomnia. Mild enough for children and elders.

LIQUORICE

Good for heartburn, irritable bowel syndrome (IBS), sore throat, coughs and tendonitis.

NETTLES

Packed with vitamins, minerals and chlorophyll, nettles help with tiredness and tension.

MARSHMALLOW ROOT

Recommended for whooping cough and bronchitis.

ROSE

Pink and red rose petals are high in bioflavonoids, which are antioxidant and anti-inflammatory.

TURMERIC

A magical anti-inflammatory and antioxidant. Great for arthritis, pain and achey muscles. On its own, turmeric is poorly absorbed in the bloodstream, and needs to be blended with something like black pepper to be effective.

VALERIAN

Well known for insomnia and perfect too for menstrual cramps, headaches, aches and anxiety.

VIOLET

Can be used as a blood cleanser and is good for chest infections and dry coughs. There are hundreds of species of violets worldwide, so check the Latin name to make sure it's edible before consuming.

WHITE SAGE

Good for easing congestion and respiratory issues. Also good for colds and flu.

YARROW

Anti-microbial and anti-inflammatory yarrow also helps with bruising, thread or spider veins and haemorrhoids.

Flower Essence

It is said that flower essences work through the acupuncture meridians to enhance your state of mind. I like to use them as a pick-me-up or indeed when I am having a moment of anxiety or panic – just before an exam or public speaking seems to work a treat. Be open minded when reading this and even more open when creating your own for best results.

Flower essences are similar to tinctures in that they use alcohol, but different because their creation involves a more holistic method (the sun) and flowers instead of herbs. Flower essences need to be made on sunny days as it is important for the solarisation process by using light to extract the flowers' benefits. It may seem like a hippyish process as for the most part it is about intuition and being drawn to a particular flower but at the end of the day most of these flowers have some kind of benefits anyway. Better still, even if the flower you are drawn to is not edible, you can still use it to make flower essence as no trace of the plant will be found in the finished essence.

Here I describe the way I was taught to make it in my herbal medicine course, but herbalists and naturopaths may have slightly different takes on this method, and you will see online that people mainly use a similar base recipe with slight variations. Have a go and see if it works for you.

Go for a walk, pick some flowers. Or go to the market, buy some flowers. However you end up with them, choose the ones you are drawn to or those that have the health associations you are after. Now choose a small bowl, fill it with spring water and put it in the sun, preferably among the flowers the essence is being made from if you picked them from your garden. If not, no stress.

You need to put the least possible amount of your own energy into the essence – you want only the flowers to impact on the water. It is important to consciously distance yourself from the action. By this, I mean, when collecting the flowers, imagine you are just an instrument, try and remain neutral.

Gently drop the flowers onto the surface of the water. They don't need to be submerged, just gently dropped onto the surface of the water is perfect. Leave in the full sun (no shadows) for two hours. After that time, remove the flowers with a twig, leaf or branch. Don't use your hands (this is so you don't transfer any of your energy). Sounds weird, just go with it.

Using a glass funnel, pour the water into a sterilised glass bottle until it is half full. Fill the bottle to the top with the brandy or vodka. This becomes the 'mother essence' or 'mother tincture'.

Take your bowl inside and take seven drops from the mother tincture and put them into a small (20ml) bottle. Fill the rest of the bottle with alcohol. This is now your stock bottle: any medicine is made from this bottle. Put 4–7 drops into a small dropper bottle and fill with alcohol. Label the bottle. This is now your flower essence.

To use, add a couple of drops to your water or directly under your tongue.

Everyday Tonic for Health

Take this daily as a tablespoon or shot. At first it will be a little hard to handle, but persevere and after a week or so your body will crave the acidity and in turn keep the benefits from all of the incredible herbs and spices.

MAKES 300ML

3 garlic cloves, peeled
1cm piece of fresh ginger, peeled
2.5cm piece of fresh
   turmeric, peeled
1 lemon, peeled
1 teaspoon horseradish
pinch of cayenne pepper
2 tablespoons raw honey
300ml organic raw apple
   cider vinegar

Put the garlic, ginger, turmeric, lemon and horseradish through a juice extractor, then mix in the other ingredients. If you don't have a juicer, grate the garlic and all the roots, squeeze in the lemon juice and some zest and mix with the other ingredients.

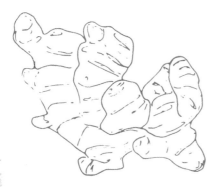

garlic and Lemon Tonic

This will do two things for you. First it will keep the flu away and second, it will keep your loved one from kissing you. No, in all seriousness, this is jam-packed full of goodness and all you need to do is take a shot a day during flu season or when you are feeling run down.

Put the whole lemons, peel and all, with the garlic in a Vitamix blender and process until smooth. Add to a medium-sized saucepan with the water and bring to the boil. As soon as it's boiling, remove from the heat and once cool, strain into a sterilised bottle. Keep refrigerated and take one shot per day until finished, or when you are feeling run down. It will last up to a year.

MAKES JUST OVER 300ML
5 lemons
30 garlic cloves, peeled
1.5 litres water

sterilised bottle

Garlic is not just famed for its culinary uses, it has long been celebrated for its medicinal benefits too. It is an excellent source of minerals and vitamins. The bulbs themselves are a rich source of potassium, iron calcium, magnesium, manganese, zinc and selenium, which is great for the heart.

Medicinal Honey (Cough Syrup)

Oh, honey honey. Honey has so many properties that keep us healthy – from antibacterial to anti-inflammatory. When the season turns cold, the dreaded coughs and colds begin. A sore throat can mean sleepless nights and one of the simplest solutions is honey. It acts in a similar way to cough syrup by coating the throat and soothing it. Add herbs to it and you have your very own natural cough syrup. Use it in your teas as a herbal sweetener or take a tablespoon before bed or when necessary. My cousin Ashleigh and I recently did a beekeeping course and we just painted our bee hives sage green. We can't wait for our bees and their queen to arrive next season so we have honey that's as local as it gets.

MAKES 300–400G

I would suggest any of the following to ease a cough or sore throat:

elderberry
hawthorn berries
angelica root
turmeric and black peppercorns
   (just a few)
lemon balm and lemon verbena
   (a mix of the two)
calendula
rose petals
kakadu plum (gubinge)

You can use bought or homegrown and dried herbs, roots, flowers and spices. Just make sure they are ground reasonably finely in your pestle and mortar or food processor. Mix equal proportions of honey with herbs (I suggest you use 200g honey to 200g herbs for this recipe). Put all of the ingredients into a double boiler and keep over low heat for 6 hours, making sure that the honey doesn't exceed 43–46°C. If you don't have a double boiler you can easily improvise by placing a smaller pan inside a larger one, using metal jar lid rings or a small round wire cooling rack to keep the inner pan off the bottom of the larger pan. Fill the larger pan with hot water to come halfway up the sides of the smaller pan. Heat gently; you may need to add more water to the pan due to evaporation. Make sure your honey completely covers the herbs, adding more if necessary. Stir the mix every now and then so it infuses evenly.

After 6 hours, strain the honey while it is still warm, using a muslin cloth, or a tighter-weave cheesecloth, even a clean cotton T-shirt is fine. Press the mix through the material, wringing it until it comes through. Pour into a sterilised jar and keep for up to a year.

You can either use shop-bought or homegrown and dried herbs — just make sure they are ground reasonably finely. For the best health benefits, use local honey. This means the bees have foraged on local flora, which can help with respiratory issues.

Bath Salt Remedies

Bath salts not only add a beautiful aroma to your bath but they are therapeutic too. You can use many different salts, all of which have different benefits as do the individual oils. Here are a few suggestions for different baths, depending on your mood and needs. I mix magnesium and Epsom salts together as they make a great combination. The magnesium helps to keep the nerves and muscles in order and regulates the heart while strengthening bones. Epsom salts, which are a magnesium sulphate, are great for detoxifying the body.

SLEEPY TIME (MAKES 450G)

300g magnesium salts

150g Epsom salts

20 drops lavender essential oil

ENERGY (MAKES 450G)

300g magnesium salts

150g Epsom salts

10 drops peppermint essential oil

10 drops rosemary essential oil

ACHEY BODY (MAKES 450G)

300g magnesium salts

150g Epsom salts

10 drops eucalyptus oil

10 drops lavender essential oil

BUSY DAY AHEAD (MAKES 450G)

300g magnesium salts

150g Epsom salts

15 drops sweet orange essential oil

10 drops lemon essential oil

Sleepy Time Massage Oil

I used to have insomnia. It is apparently hereditary and both my grannies have had it. I had it for four years. When my doctor diagnosed me with insomnia, I was given a prescription but never took the pills. I knew for me it would have been a slippery slope and instead tried to fight the condition without medication. It was tough, I am not going to lie. Then I moved out of London to rural England when I was accepted onto my masters course in sustainable agriculture. It was like a miracle – on my first night out in the country, I slept like a baby. To me this says that sometimes (not in every case, of course) but sometimes it is down to environment and mental stimulation. A massage with this oil and some sleepy tea sends me right back to rural England where my best years of sleep were ever had.

MAKES 250ML

120ml coconut oil

120ml grapeseed or sunflower oil

15 drops orange essential oil

10 drops lavender essential oil

5 drops ylang ylang essential oil

5 drops geranium essential oil

5 drops rose essential oil

Pour the oils into a clean dark-glass bottle using a small funnel. Add the essential oils. Screw on the lid and shake. Store in a cool, dark place for up to a year.

Burn Salve

Nobody likes a burn and this recipe truly works to soothe minor burns. The lavender oil helps to calm the inflamed injury all the while disinfecting it. The oil provides vitamin E to soothe the skin with nutrients, coconut oil helps fight bacteria, while honey provides antioxidants and rehydrates the poor burn, helping to stop it from blistering.

MAKES 30ML

30ml honey (manuka if possible)
2 tablespoons extra virgin olive oil
1 tablespoon coconut oil
20 drops lavender essential oil

small sterilised glass jar

Mix all the ingredients together and store in the jar. When needed, spread lavishly over the burn. Cover the injury site with a dressing or plaster.

This will keep for up to a year.

Burns can be serious. If this does not help to ease the pain, seek medical attention.

Sting goes

Insect bites and stings can be annoying and painful, to say the least. All they need is a little something to ease the inflammation and stop any infection or bad bacteria. These few things combined do just that.

MAKES 1 SMALL TIN

3 teaspoons beeswax
1 tablespoon coconut oil
4 drops lavender essential oil
3 drops eucalyptus essential oil
½ teaspoon raw honey

Melt your beeswax and coconut oil together in a small bowl set over a pan of simmering water, then remove from the heat. Stir in the essential oils and honey. Pour into a small tin, put the lid on, and let it cool completely before use. This will keep for up to a year.

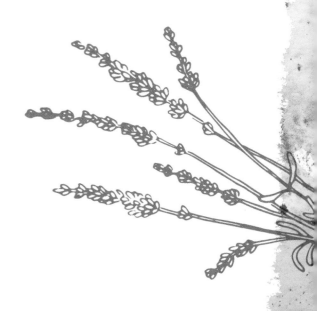

Anti Inflam tea

Perfect for both achey bodies and period pain. The ingredients blended together will help reduce inflammation that comes from being on your feet, sport, arthritis and that dreadful time of the month (insert sad face here). Take as needed.

Blend and store in an airtight container for up to a year. Use approximately 3g per mug of boiling water.

MAKES 140G

30g dried lemon balm
30g dried lemon verbena
30g dried hibiscus
30g dried calendula
10g dried meadowsweet
10g dried rose petals

This also makes a wonderful iced tea. Prepare as above and add a teaspoon of raw honey, then mix until dissolved. Let the tea cool to room temperature and serve over ice with some fresh herbs.

Herbal Tinctures

I just love the word tincture. It makes me feel like an olden-day alchemist. I would love a dispensary where I can dispense tonics and tinctures for people's health. There would be big jars filled with dried herbs and bottles with handwritten labels. Herbal tinctures are basically concentrated herbal extracts, and they are another way to benefit from the healing powers of herbs. Alcohol acts as the solvent to draw out the nutrients from the herbs. You can use any alcohol you like but here I'm using vodka and pure grain alcohol (PGA). You don't need much: you can just drink a spoonful or dilute it in water or juice. You can use fresh or dried flowers. Use the herb guide on pages 174–5 to see which herbs will give you most benefit and then follow the method below.

EITHER:

80–90 per cent proof vodka – good for dried herbs and some fresh hard herbs

OR:

a combination (50:50) of 80 per cent proof vodka mixed with 190 per cent proof grain alcohol – good for high-moisture herbs dried or fresh herbs or roots and bark

sterilised jars with screw-top lids
a dark glass bottle

To make a tincture using dried herbs, finely chop enough herbs to three-quarters fill whatever size jar you have. Or, if you are using fresh herbs (or flowers, roots or bark), very finely chop enough to fill the jar half full. Ensure the jar is thoroughly clean and do chop the herbs really fine. Pour in the alcohol to fill the jar right to the top. Screw on the lid and store in a cool, dry place away from sunlight.

Store your tincture in a cool, dry place with no sunlight. Shake several times a week and check that the alcohol has not evaporated. If the herbs are not covered by the alcohol, you must top up the jar – if they are exposed to air you risk getting mould and bacteria into your tincture. Allow the mixture to extract for 8 weeks.

After that time, line a funnel with a piece of cheesecloth and set over the dark glass bottle. Pour the contents of the jar into the funnel, squeezing the herbs to extract as much tincture as possible. Seal, label and date the bottle. Kept in a cool, dark place your tincture will last for many years. Try to have a tablespoon a day if possible but you may need to start by mixing it in water, as it's very potent if you aren't used to it.

Resources

It's always great to try and buy your fresh ingredients locally, but here are some of the suppliers I like to use.

UK
PESTLE HERBS
Dried herbs and apothecary bottles
Delivery within the UK and Northern Ireland
www.pestleherbs.co.uk

BALDWINS
Dried herbs, candle and soap equipment, essential oils, carrier oils (everything you need)
Delivery worldwide
www.baldwins.co.uk

AMAZON
A huge range of ingredients and equipment.
www.amazon.co.uk

AUSTRALIA
NEW DIRECTIONS
Cosmetic bases, essential oils, castile soap, bottles and jars, packaging and dried ingredients
Delivery worldwide
www.newdirections.com.au

AUSTRAL HERBS
For dried herbs and essential oils
Delivery within Australia
www.australherbs.com.au

ESSENTIALLY AUSTRALIA
Essential Oils
Delivery within Australia
www.essentiallyaustralia.com.au

WHOLESALE MINERAL MAKE-UP
Mica and make-up ingredients
Delivery worldwide
www.wholesalemineralmakeup.com.au

USA & CANADA
MOUNTAIN ROSE HERBS
Delivery within USA and Canada
www.mountainroseherbs.com

SPECIALTY BOTTLE
Bottles and cosmetic equipment
Delivery worldwide
www.specialtybottle.com

ORGANIC ALCOHOL
Alcohol for tinctures and tonics
Delivery worldwide
www.organicalcohol.com

AMAZON
A huge range of ingredients and equipment.
www.amazon.com

FURTHER READING
www.swsbm.com/ManualsMM/MatMed5.pdf

Beentje, H. Kew Plant Glossary: An Illustrated Dictionary of Plant Terms (London: Royal Botanic Gardens, 2015).

Castner, James L. Photographic Atlas of Botany and Guide to Plant Identication (Gainesville, FL: Feline Press, 2004).

Elpel, T. J. Botany in a Day: Patterns Method of Plant Identification (Pony, MT: HOPS Press, 2004).

Harris, J. G., and M. W. Harris. Plant Identification Terminology: An Illustrated Glossary (Payson, UT: Spring Lake Publishing, 2001).

Heywood, V. H. Flowering Plants of the World (Oxford University Press, 1993).

Index

A

aches and pains 167, 182, 184–5
air fresheners 14
aloe vera gel 154–5, 158
anti inflam tea 184–5
anti-aging face serum 126
antibiotics 168
apple, pear and tomato
  chutney 76
apple cider vinegar, raw
  in cosmetics 128, 148, 151
  in culinary recipes 49, 74–6,
    78–9, 98
  in household products 20, 27
  in remedies 168, 170, 178

B

bacon, homemade 56–7
balm, achey breaky 167
banana and cinnamon scrub 142
basil 111, 152
basil and lemon thyme smash
  112
bath fizzes 138–9
bath salts 136–7, 182
bathroom cleaners 30–1
beard oil 164
beeswax 158, 167, 183
beetroot 92
beetroot chutney 74
berries
  berry cordial 98
  berry jam 84
  see also specific berries
bicarbonate of soda
  in cosmetics 136, 138, 144–5
  in household products 21, 24,
    27, 28, 30, 31
bird feeders 118–19

bitters, pink grapefruit 99
blackberry and pear sorbet 91
blueberry 92
blueberry and honey mask
  132–3
blusher 156–7
body butters 134, 167
body scrubs 142–4
bone broth 54–5
bug motels 119
burn salve 183
butter
fruit butter 85
making dairy butter 44–5
nut butters 41
buttermilk pickles 78

C

cabbage 70–1, 73
cacao powder 154–6
calendula 174, 184
candles 16–17, 115
cardamon and rose hot choc 104
carpet cleaner 26
carrot and ginger kraut 73
carrot top pesto 46
cherry, pickled 79
chocolate
  choc coffee man scrub 165
  floral hot chocolate 104–5
chutneys 74–7
cinnamon 26, 145
banana and cinnamon scrub 142
citroncello 102
citroncello spritz 113
citrus
  citrus all-round cleaning spray
    28
  citrus curd 85, 86
  see also lemon; lime; orange
cleaning products 6–7, 12–31
cocktails 110–13

cocoa/cacao butter 134, 165,
  167
coconut
  coconut, honey and mint face
    and body scrub 142
  coconut conditioner 148–9
  coconut and orange shampoo
    148–9
  coconut shaving cream 164
  coconut oil 126
  coconut oil hair mask 150
  and health and beauty 141–2,
    144–5, 148, 150, 152, 155,
    158, 165
  and male grooming 164, 165,
    167
  and remedies 182–3
coffee grinds 114–15
choc coffee man scrub 165
compost 115
concealer 154
conditioner, coconut 148–9
cordial, berry 98
cough syrup 180–1
courgette, pickled 78
crackers, veggie scrap 68–9
curd, citrus 85, 86
cured fish 60
curry
  green curry paste 61
  red curry paste 62–3

D

dandruff 148, 152
deodorant 144
detanglers 151
drinks trolley 94–105, 110–13

E

eggs, and hair care 150–1
energy bath salts 182
Epsom salts 24, 136, 138, 144,

182

essential oils
and health and beauty 138,
144, 152
household products 19, 22,
24, 28, 30
see also specific oils
eucalyptus essential oil 26, 141,
167, 182–3
eye make-up removers 128

F

fabric softener 24
face masks 130, 132–3
facial hair 164
fermentation 70–3
fig and peach chutney 74
fish, cured 60
flowers
flower essences 176
pot pourri 18–19
soap 22
food colourings, natural 92–3,
138
foot scrubs 144
foraging 7

G

garden, the 106–121
garlic 174, 178
garlic and lemon tonic 179
ginger 141, 174, 178
carrot and ginger kraut 73
ginger beer 101
lemongrass, lime and ginger air
freshener 14
green tea and spirulina face mask
130
gut flora 71, 96–7

H

hair care 115, 146, 148, 150–2,
162
health and beauty 123–45
herbs
dried 34, 168
freezing in oil 34, 36
guide 174–5
herb-infused oils 64
herb-infused vinegar 64
and household products 22, 26
see also specific herbs
hibiscus 99, 174, 184
holistic approach 6
homes 11–31
honey 126, 165, 178, 183
blueberry and honey mask
132–3
coconut, honey and mint face
and body scrub 142
health benefits 168
honey hair mask 150
honey and rose facial toner 128
mango, lime and honey sorbet
91
medicinal honey 180–1
sage and honey spritz 112

I

ice cream, vanilla 88–9
ice cubes, no waste 36–7
infusions 62–3, 64–5

J

jam 84
jojoba essential oil 152, 154, 164

K

kimchi 73
kombucha 96–7

L

laundry detergent 24
lavender 14, 26, 104, 111
sage and lavender hot
chocolate 104
lavender essential oil
and health and beauty 130,
136, 141, 144, 148, 151–2,
162, 164
household uses 24, 31, 136
remedies 182, 183
lemon
culinary uses 75, 82–3, 85–6,
102, 112
garlic and lemon tonic 179
and health and beauty 150, 151
herby floral lemonade 102
lemon hair mask 150
preserved lemons 75, 82–3
use in remedies 178–9
lemon balm 111, 175, 184
lemon balm mint vodka
muddle 113
lemon balm sugar 113
lemon essential oil
for health and beauty 130,
136, 142, 182
household uses 24, 31
rosemary and lemon bath salt
soak 136
lemon thyme 111
basil and lemon thyme smash 112
lemon verbena 175, 184
lemongrass 175
lemongrass, lime and ginger air
freshener 14
lime 85
lemongrass, lime and ginger air
freshener 14
mango, lime and honey sorbet 91
lip balm, rose 158
local produce 6

M

macadamia oil 167
macadamia and vanilla body
  butter 134–5
magnesium salts 182
make-up 146, 154–9
male grooming 160–7
mango, lime and honey sorbet
  91
mascara 158
masks
  face 130, 132–3
  hair 150
massage oil 140–1, 182
mica powder 155–6, 158
mint 26, 108, 111
  coconut, honey and mint face
    and body scrub 142
  lemon balm mint vodka
    muddle 113
  see also peppermint essential oil
moisturisers 134, 155, 167
moths be gone 26
mouthwash 145

N

nettle 162, 175
nut milks 40
nut oils 126

O

oils 126
beard 164
infused 62–3
massage 140–1, 182
see also specific oils
olive oil 20, 150–1, 164, 183
olive tapenade 46
onion, spicy pickled 80–1
oral hygiene 145
orange essential oil
  coconut and orange shampoo 148

and health and beauty 130,
  142, 148
and household products 20,
  24, 31
and remedies 182
oven and hob cleaning 27

P

pasta, homemade fresh 50–3
peach and fig chutney 74
pear
  apple, pear and tomato
    chutney 76
  plum and preserved lemon
    chutney 75
  spiced pear and blackberry
    sorbet 91
peppermint essential oil 141,
  144–5, 167, 182
  lavender and peppermint bath
    soak 136
period pain 184
pesto, carrot top 46
pickles 78–81
pink grapefruit bitters 99
plum
  plum, pear and preserved
    lemon chutney 75
  plum jam 84
polish, wood 20
pork, potted 56–7
pot pourri 18–19
probiotics 71, 96–7

R

remedies 168–86
room sprays 14–15
rose 175, 184
  honey and rose facial toner 128
  rose cardamon hot choc 104
  rose essential oil 31, 182
  rose lip balm 158

rosemary 26, 111, 142
  rosemary, sage and lavender air
    freshener 14
  rosemary and lemon bath salt
    soak 136
  rosemary essential oil 152, 162,
    182

S

sage 111, 148
  rosemary, sage and lavender air
    freshener 14
  sage and honey spritz 112
  sage lavender hot choc 104
  see also white sage essential oil
salad dressing 108
salad garden, miniature 108–9
salt(s) 126
  bath salts 136–7
  infused salts 38–9
  non-iodised salts 31
  salt hair spray 152
  salt scrubs 142, 144
sandalwood essential oil 164,
  167
sauerkraut 70–1
scalps, itchy sore 152
SCOBY 96–7
scrubs
  body 115, 142–3, 165
  facial 115, 130, 165
  foot 144
  scum 31
scum scrub 31
seasonal produce 6
serum, anti-aging facial 126
shampoo, coconut and orange
  148–9
shaving cream, coconut 164
shea butter 134, 154, 158, 164,
  167
shrub (drink) 98

silicones 146

skin care 6, 115
  bath products 136–9
  body care 134, 142–4, 155, 165
  facial care 126, 128, 130, 132–4, 155

sleepy aids 182

smells, absorption 115

soap
  Castile 28, 144, 148, 164
  home-made 22–3
  for washing powder 24

sorbet, pure fruit 90–1

spirulina and green tea face mask 130

spritzes 112, 113

sting goes 183

storecupboard essentials 8

stove top cleaner 27

strawberry super sorbet 91

sugar 126, 130, 132, 142, 165
  lemon balm sugar 113
  sugar face scrub 130

sun-protection factor (SPF) 155

sustainability 6, 7

T

tapenade 46

tea
  anti inflam 184–5
  kombucha 96

tea tree essential oil 145, 148, 152

terrariums 116–17

tinctures 168, 170

toilet bliss bombs 31

tomato
  apple, pear and tomato chutney 76
  sweet and spicy tomato sauce 48–9

toner, honey and rose 128

tonics, every day, for health 178

toothpaste 145

turmeric 92, 175, 178

V

vanilla
  macadamia and vanilla body butter 134
  vanilla ice cream 88–9

veggie scrap crackers 68–9

vinegar
  balsamic 74
  brown malt 80–1
  in household products 20, 26–8
  infused 62–3
  medicinal 170
  red wine 75
  white 20, 26, 28
  white wine 79, 80–1
  see also apple cider vinegar, raw

vodka 102, 170
  lemon balm mint vodka muddle 113

W

washing powder 24

washing soda 31

washing-up liquid 28

white sage essential oil 162, 175

wildlife 119

wood polish 20

wreaths 120–1

Acknowledgements

So many people to thank and so little space. Firstly to my family. The small one and the big one. My mum and dad have always supported me in whatever path I have taken and I love that they sit in what my dad calls his 'proud chair' because of the path I am on. All I have ever wanted was to make them and my brothers proud. So to you my small family and the rest of my big extended family, especially its oh-so-wonderful leader, my nan. To all of my uncles and aunties, Sarah, Paul, Mark, Kylie and Angie (and Paul's love Bec and Nan's love Harry). All of my cousins, Yasmin, Sam, Teryn, Ashleigh, Caitlyn, Liam, Brad and Taylah and then the rest of our little family Emma, Koen and my godson Charlie. You are all everything to me. As are you Damien, my love. Thank you for putting up with our home looking like a constant laboratory and test kitchen and being my guinea pig. To my friends who have supported me for the decades, who have loved me no matter what and pushed me to always be the best version of myself.

To Kyle: thank you so very much for taking a chance on me. Thank you from the bottom of my heart for allowing me to be me. Here is to a long journey making beautiful and meaningful books together. To my team. The A team. Tara, the most incredible Editor a girl could ask for. Your patience, generosity and passion for this book made it what it is. Nassima – wow, what a force you are girl, both behind the lens and not. Thank you for making my recipes and creations come to life and reading my mind most of the time. I only hope we work on many more things together. Rachel, no words can thank you enough for being the most incredible partner in crime styling this book and seeing inside my messy brain. You are so immensely talented and I am so grateful to not just have you as a friend but to have worked with you on this. Thank you also to Jen for the fantastic props, and to the great assistants on the photo shoots, Jack, James, Celia and Emma. Thank you to Laura for your beautiful work on the book design.

To all of the people in my work world (some of whom are now dear friends too) who have taught me so very much over the years, in particular Valli Little and Costa Geogordis, who have become my benchmarks of authenticity and integrity. Thank you. There is no way I would be where I am without you teaching me everything I know. Last but not at all least, to all of you who bought this book. Massive gratitude from the bottom of my heart.